THE WORKBOOK OF Improving Maintenance & Reliability Through Cultural Change

Stephen J. Thomas

Industrial Press

Industrial Press, Inc.
1 Chestnut Street
South Norwalk, CT 06854

Tel: 203-956-5593, Toll-free: 888-528-7852

Email: info@industrialpress.com

Websites: books.industrialpress.com
ebooks.industrialpress.com

First Edition, January 2006

Sponsoring Editor: John Carleo
Interior Text and Cover Design: Janet Romano
Developmental Editor: Robert Weinstein

Table of Contents

Chapter 1	The Beginning	1
Chapter 2	Culture Defined	13
Chapter 3	Vision and the Goal Achievement Model	29
Chapter 4	Organizational Values	47
Chapter 5	Role Models	77
Chapter 6	Rites and Rituals	103
Chapter 7	Cultural Infrastructure	125
Chapter 8	The Elements of Change	149
Chapter 9	Leadership	165
Chapter 10	Work Process	191
Chapter 11	Structure	215
Chapter 12	Group Learning	235
Chapter 13	Technology	261
Chapter 14	Communication	283
Chapter 15	Interrelationships	303
Chapter 16	Rewards	323
Chapter 17	The Web of Cultural Change	343
Chapter 18	Assessment and Corrective Action	361
Chapter 19	Moving Forward	375
Appendix 1	The American Tractium Production Co.	385

Dedication

This workbook is dedicated to my daughter Allison (Allie) Thomas.
If you define your goals, have confidence in your ability
and never give up there is absolutely no limit to what you can achieve.

Acknowledgments

I would like to thank Industrial Press, Inc especially John Carleo for once again having faith in my work and ideas and helping me to convert them into a workbook, and to Janet Romano for her hard work to make this a quality product.

I would also like to thank Robert Weinstein for his support and editing skills that helped me to be focused and deliver a product of which I can be proud.

Last, but not least I would like to thank my wife Susan for her support and patience as I work in my office.

The Beginning

Chapter 1

Purpose

This chapter introduces the reader to the workbook and how to use it. It discusses breaking down learning barriers and introduces the web models for the eight elements of change and four elements of culture. The chapter concludes with an explanation of alternative ways that the material can be learned or utilized when developing a training program.

Contents

Introduction

Benefits

The Subjective vs. Objective Barrier

Other Barriers

The Eight Elements of Change and the Four Elements of Culture

Figure 1-1 The Eight Elements of Change and Their Effect on One of the Four Elements of Culture

Figure 1-2 The Four Elements of Culture and Their Effect on One of the Eight Elements of Change

Exercise 1-1 Overcoming Barriers

Figure 1-3 Machiavelli Quotation

Exercise 1-2 Recognizing a Cultural Problem

How To Use This Workbook

Figure 1-4 Book Navigation Diagram

Introduction

One of the concepts presented in *Improving Maintenance and Reliability Through Cultural Change* is how hard it is to change the culture of an organization. The quote in Figure 1-3 by Nicolo Machiavelli, written over 500 years ago, clearly points out the enormity of this task. Change is difficult for the simplest of tasks that are culturally ingrained and even more so for the more prevalent work processes such as plant maintenance and reliability. The question then is how do we make these cultural shifts in our organizations?

I hope that, having read the book, you have discovered new information that you wish to share with your organization, ultimately leading it to new and better ways of doing things. Although this idea is a very noble one, it will be difficult, to say the least. However, this workbook can help. The exercises contained within these pages will help you assist those within your organization who need to recognize and make the shift, enabling them to understand and ultimately embrace these new and sometimes different concepts.

The workbook has been created so that each chapter corresponds to the matching chapter in the book. This structure will be helpful if you follow the book format when developing your training plan. It will also be helpful if you want to provide specific chapter-focused information; the workbook and its related chapter in the book can be used as a single training module.

Benefits

Each workbook chapter has several parts that will provide you with value as you roll out the changes and associated training at your plant. First are the full-page diagrams. These diagrams are the same as those in the book. They have been placed in the workbook so that you can copy them and hand them out to your team members. In this fashion, you can use them to make specific points as you address the issues to which they relate. The diagrams can also be easily scanned and used as slides to provide visible support or discussion points for your effort.

Second are the individual exercises. These have been developed to help you better understand a concept being presented in the book. They are designed to be worked on independently so that the acquired knowledge from completing them will be of personal benefit to you. An alternative is to complete them in a group setting with associated discussion so that you and your group or work team can achieve the knowledge sought.

Third are the group exercises. These are specifically designed for a group setting. Learning something as an individual is often not nearly as significant as learning it in a group setting. Groups working through the material can collectively realize the value contained in the exercise and reinforce this within the group. These group exercises will also work well in a training setting because the instructor can use them to get participants out of their seats and interacting. These exercises therefore stimulate networking which, in turn, allows for the transfer of knowledge among the group members.

The fourth element of the workbook is the case studies. These are simulations of a live plant and the events related to change. By using these case studies, the participants whether as individuals or in a group setting can apply what they have learned to a simulated plant situation. Through case study work, what has been learned throughout other types of exercises and in the text can now be applied. This powerful reinforcement of what was learned can go a long way towards making the application of the newly-learned skills easier when the student returns to the real world.

The open-ended questions are the fifth element of the workbook. They are included in many exercises. They can be worked on individually or in a group setting. These questions have no right answer. They are designed to promote individual thought and, where applicable, group discussion. In a group format, these exercises allow people with different ideas and backgrounds to discuss the material with others who may not hold the same ideas.

When all of these elements are combined, they create a very effective training program, one that will go a long way towards promoting cultural change in your organization. Because internally-developed training programs are unique and usually address specific needs, the last section of this chapter (and Section 1.6 of the book) have been created to show various ways to navigate through the material and, if desired, build a training program that fits your needs.

The Subjective vs. Objective Barrier

All too often, when people read books or use workbooks that ask for opinions, either on a group or personal level, the comment is often heard "this is only subjective information." This comment implies that subjective data does not have as much value as objective data simply because it is someone's opinion. This barrier is thrown in the way of learning how to change your organization. It is a formidable barrier if these words are spoken by your boss or any member of senior management whose support you need to make the change. However, subjective information should not be dismissed so lightly.

There is an old saying that "when you have a feeling about something, go with your feelings because they usually will be right." This concept is even truer when dealing with a group or, in this case, your learning team. Consultants who go into a business, first assessing the work process, then recommending improvements, essentially base many of their recommendations on subjective opinion. They interview many people, asking them to identify problems with the work process. The consultants use this information to reach conclusions regarding potential areas for improvement. They also seek supporting objective data, but much of their basic work revolves around subjective group opinion.

Many of the exercises in this workbook are based on subjective opinion. Trust me; this information is valid and valuable.

Other Barriers

Other barriers to change exist in addition to the subjective–objective one just discussed. These barriers are used for a number of reasons to maintain the status quo. First, the individuals presenting the barrier may actually think that the status quo is the right thing for the business. They may see the needs of the organization far differently and may not see a need for change. Another reason may be that they are afraid of the risk associated with changing the current practices. They may believe that the negative implications of failure are far more risky that the positives that could emerge from a successful change effort. They may also believe that the organization is doing well under the present processes; therefore, why bother to rock the boat? Finally they may be so transactional (day-to-day focused) that they completely fail to see how doing things differently could add any value. Exercise 1-1 will help you think about barriers and how to work through them.

The Eight Elements of Change and the Four Elements of Culture

The eight elements of change and the four elements of culture are not independent building blocks within the process of change. Quite the contrary. They each have significant influence on one another. This relationship will be demonstrated throughout this workbook (see also book page 110).

Chapters 4 through 7 each address a separate element of culture. In each, the learner will complete a portion of the Cultural Web of Change survey, answering the questions for each of the eight elements of change that pertain to that element of culture. The result will be a web for that cultural element with radial spokes being the eight elements of change. This information will enable the learner to examine each cultural element as it relates to the elements of change. The web diagram is shown in Figure 1-1.

Chapters 9 through 16 each address a separate element of change. In each, the learner will conduct a portion of the Cultural Web of Change survey, answering the questions for each of the four elements of culture related to the specific element of change. The result will be a web for that change element with radial spokes being the four elements of culture. This information will enable the learner to examine each change element as it relates to the elements of culture. The web diagram is shown in Figure 1-2.

Figure 1-1
The EightElements of Change and Their Effect on One of the Four Elements of Culture

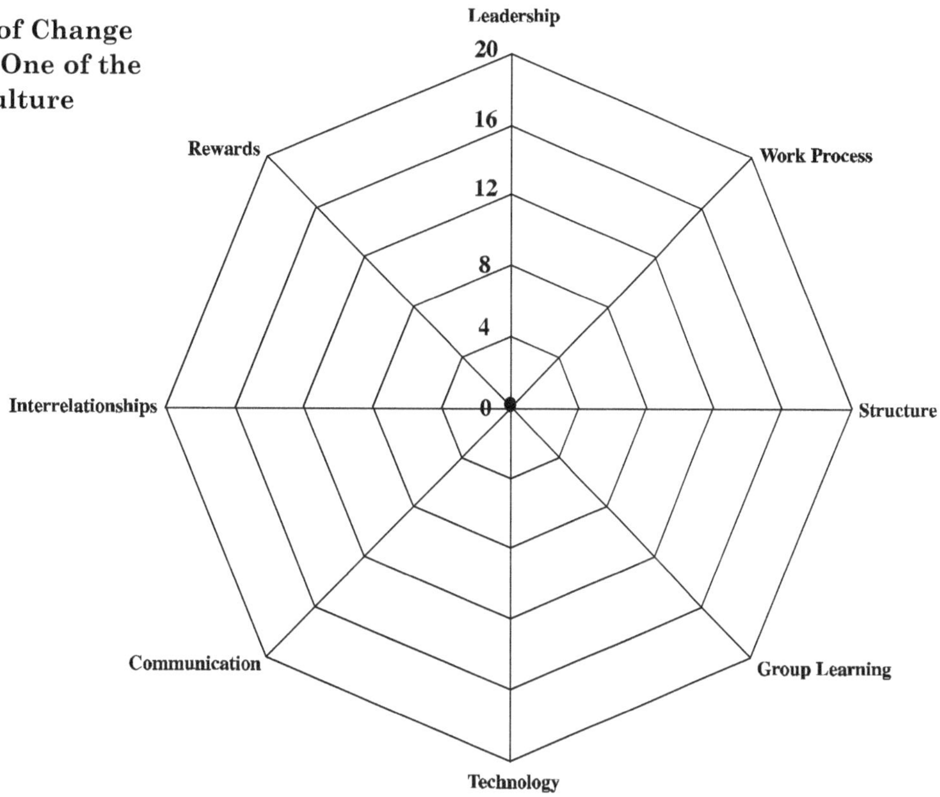

Figure 1-2
The Four Elements of Culture and Their Effect on One of the Eight Elements of Change

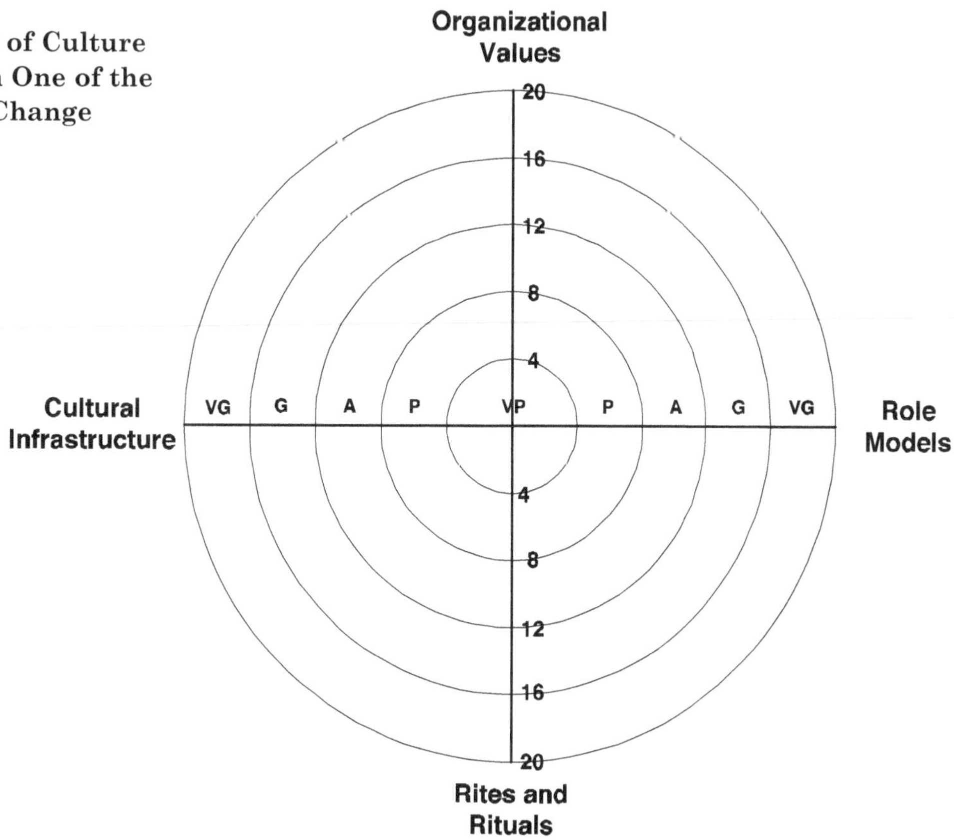

Exercise 1-1 Overcoming Barriers

What would you do? You have an idea that would drastically change the way that maintenance work planning, scheduling, and field execution would be handled in your plant. Your idea calls for dedicated planners in each of the three maintenance areas. It would be their job to develop detailed plans ready for execution. A dedicated work scheduler would then work with the three area production supervisors and develop a work schedule for the upcoming week. The work crews and their foremen would be centrally dispatched, going where the priority work and the schedule sends them. You came up with this idea based on your past experience with your former employer and from close observation of the current process. From these, you have concluded that the current reactive method of working under-optimizes the entire organization. Complete the table describing how you would overcome the barriers described.

Suppose that following your presentation, your manager makes one of these comments:

Manager's Comment	How Do You Overcome the Barrier?
We tried this and it didn't work before.	
We don't have time to make a change of this sort. We need to support Production.	
Why change? We are doing fine using our current process.	
Have you checked to see how others are doing this type of work? We want to emulate the best.	
Have you asked John the maintenance field execution manger who was promoted to this position for his success with the current work processhis opinion?	
This is just you opinion. Where is the data to support your conclusion?	

Figure 1-3 Machiavelli Quotation

There is nothing more difficult to take in hand,
more perilous to conduct, or more uncertain
in its success than to take the lead
in the introduction of a new order of things,
because the innovator has for enemies,
all of those who have done well
under the old conditions, and luke-warm defenders
in those who will do well under the new.

Exercise 1-2 Recognizing a Cultural Problem

It is your first day on the job as the new plant maintenance manager. You know from discussion with the previous manager who just retired that there is a 6:30 AM meeting of the maintenance planners, the work scheduler, and the foremen to discuss the daily schedule. This meeting is short and only runs 15 minutes, providing time for the foremen to get to their crew staging areas and start the work day at 7:00 AM.

When you arrive at the meeting room at 6:25, you find it empty. This was not what you expected, so you search the building trying to determine if it has been moved to a new location. The foremen and planners offices are empty, but you encounter Pat, the maintenance scheduler, already sitting at the computer. What Pat tells you is a far different story than the picture painted for you by your predecessor. The reason that no one was at the meeting is that there has not been a meeting held for over six months.

The real process at work is that Terry, the field execution manager, gets to work at 5:00 AM, talks to the late shift production crew, and then decides what maintenance is going to work on that day. The foremen and planners then follow Terry's directions. The planners' role in this process is material coordination and field support. Essentially there are no real work plans; the schedule that is prepared is worthless.

After the work day has started, you ask Terry to explain how the process works because you find what Pat has told you hard to believe. Terry tells you everything is under control. Production is happy with the process, and the plant is running fine. You are told in great detail how Terry got promoted to field execution manager by working in the exact way that the maintenance organization is working at the present time. Production likes the process and feels that their needs are addressed in a very timely manner. Terry also lets you know that all you need to do to be a success is to stay in your office, handle the administrative details for the organization just like the previous manager, and the rest will be taken care of.

Obviously you and Terry don't see eye-to-eye. You were brought in from outside the company by corporate to help correct what they saw as serious problems at the plantpoor reliability, high cost, and safety issues.

Unfortunately your discussion with Terry turns into a heated argument. At its conclusion, you decide to seek out the plant manager's guidance so that you can begin correcting the problem. By the start of your meeting, it is obvious that the plant manager not only knows about your discussion with Terry, but also agrees with Terry's approach, telling you several stories of how Terry saved production by quickly fixing broken line equipment. As you walk back to your office and introduce yourself to employees along the way, you realize that everyone on site knows about your heated discussion with Terry and your visit to the plant manager's office.

Worksheet for Exercise 1-2

You have a cultural problem. Answer the following questions.

1. What are the plant's organizational values? What do they believe success looks like? Provide examples.

2. Who does the maintenance organization look to for guidance and why? What makes this person the role model for the organization? What is your evidence?

3. What are the key activities of a day in the life of the maintenance organization? What does production do to reinforce this behavior?

4. What things are going on behind the scenes that provide evidence that there is an unofficial communication network in the plant?

Answers to the above four questions should clearly indicate the cultural issues at work in the plant. As you work through the workbook you will learn more about each one and how to address the problems they can create.

How To Use This Workbook

Referring to Figure 1-6, there are several ways to use the material in the book and workbook. My recommendation is to go from front to back without skipping any of the chapters. However, the book is divided into sections that could be addressed separately. The first section, which includes Chapters 1 through 7, focuses on vision, the Goal Achievement Model, and the four elements of culture. Chapters 8 through 16 deliver a detailed explanation of the eight elements of change. Finally, Chapters 17 through 19 tie all of the information together via the web of cultural change.

If you prefer, you could go through the material addressing the elements of culture as track 1, bypassing the elements of change, or conversely take on the elements of change and bypass the elements of culture. Either way could work, but it is my opinion that valuable learning opportunities will be missed.

Figure 1-4 Book Navigation Diagram

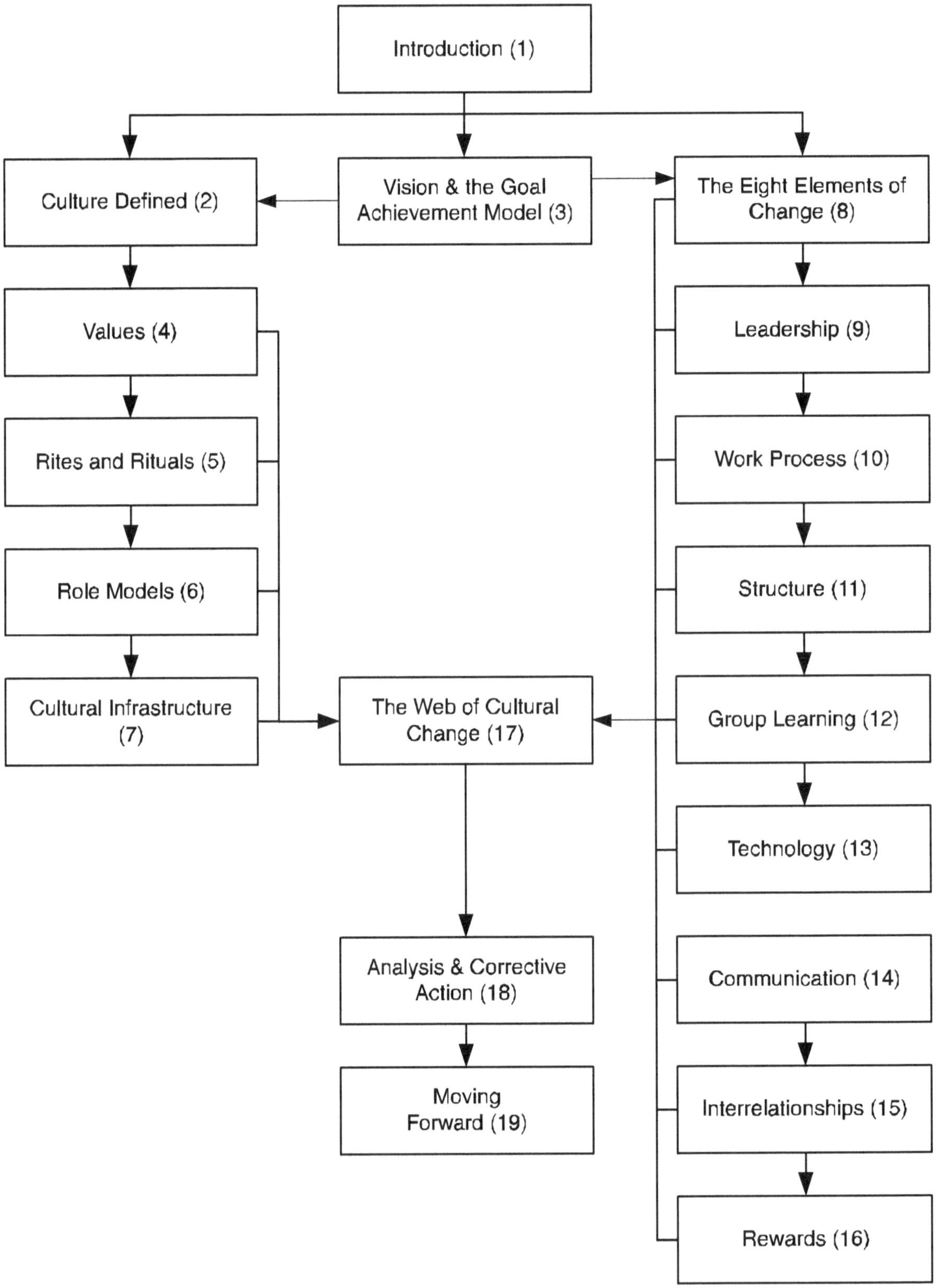

Chapter 2

Culture Defined

Purpose

This chapter introduces the subject of organizational culture. Although we can alter the eight elements of change-leadership, work process, structure, group learning, technology, communication, interrelationships, and rewards-and change the tasks we perform as part of our daily work, none of our work will be successful over the long term unless we understand and change the organizational culture.

What Is Included

Figure 2-1	Hard Skills / Soft Skills
Figure 2-2	Organizational Culture Below the Foundation
Exercise 2-1	Differentiating Between Hard Skills, Soft Skills, and Cultural Levels
Exercise 2-2	Determining the Level at Which You Need to Work
Figure 2-3	Culture Defined
Figure 2-4	The Culture Model
Figure 2-5	The Four Elements of Culture
Exercise 2-3	The Definition of Culture Dissected
Exercise 2-4	Recognizing the Parts of the Culture
Figure 2-6	How the Organization Acts Towards Change
Exercise 2-5	Organizational Culture Types
Exercise 2-6	ATPCo and Their Culture
Exercise 2-7	Your Organization's Culture

Figure 2-1 Hard Skills / Soft Skills

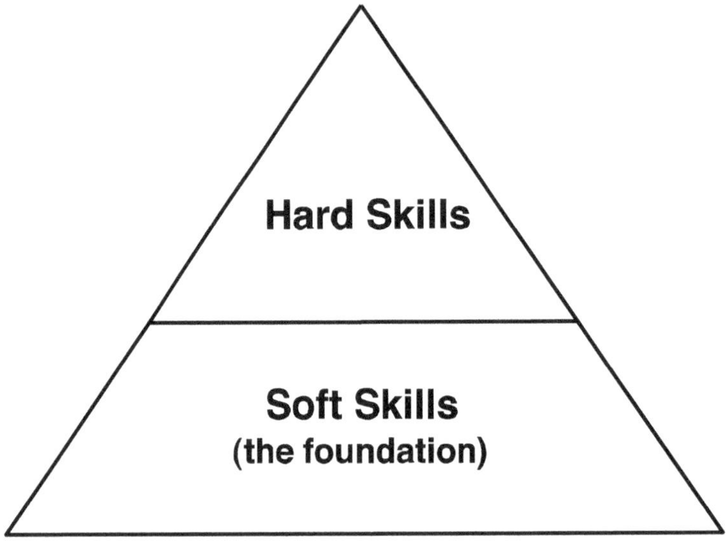

Figure 2-2 Organizational Culture Below the Foundation

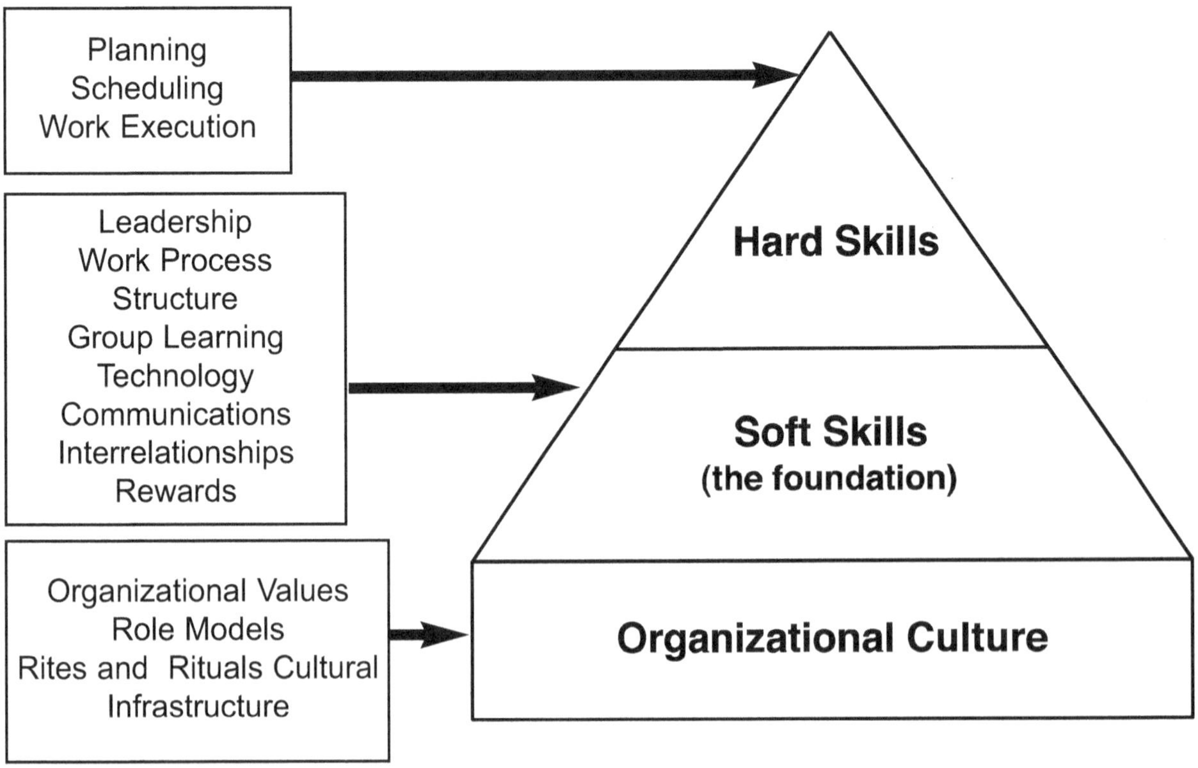

Exercise 2-1 Differentiating Between Hard Skills, Soft Skills, and Cultural Levels

Being able to differentiate between the various levels of work-hard skills, soft skills, and the elements of culture-is important. Change at the top of the pyramid is supported by first changing things at the bottom. Identifying the level at which a skills exists will help you understand the order in which the change initiatives should take place. In this exercise, identify the ten elements listed as either hard skill, soft skill, or culturally-related skill by placing the appropriate letter H (hard), S (soft), or C (culture) in the column marked Type. Then in the column marked Why explain why each element fits that category.

Skill	Type	Why
Work Planning		
Role Models		
Communication		
Leadership		
Preventive Maintenance		
Work Execution		
Rites and Rituals		
Scheduling		
Work Process		
Structure		
Reliability Improvement		
Organizational Values		
Group Learning		
Materialization		
Rewards		

Exercise 2-2 Determining the Level at which You Need to Work

Once you understand the various levels at which change-related skills exist, you need to determine where you start your work to change the status quo. This exercise lists several short problem descriptions. For each, identify the level(s) at which work is needed in order to make a value-added change. Also indicate how each level of the organization would contribute to the change effort.

Scenario	What Levels Require Work and Why?
Your preventive maintenance program is not working as well as you want and needs minor adjustment.	
Equipment failure leads to panic within the plant; everyone focuses on restoring service as quickly as possible.	
The organization's structure does not work well in the context of the new maintenance work process.	
New ideas are dismissed without investigating their potential value.	
Production, which is very interested in improved work scheduling, needs some training regarding their role in the process.	
Rewards and praise are heaped upon those who make repairs to production equipment and quickly restore service.	
You can't understand why, but every time your team makes the slightest mistake, your manager seems to know about it.	
The planning process was just initiated and it is evident that the planners need additional training.	

Figure 2-3 Culture Defined

A pattern of shared basic assumptions that the group learned as it solved its problems of external adaptation and internal integration, that has worked well enough to be considered valid and, therefore, to be taught to new members as the correct way to perceive, think, and feel in relation to those problems.

Organizational Culture and Leadership, E. Schein

Figure 2-4 The Culture Model

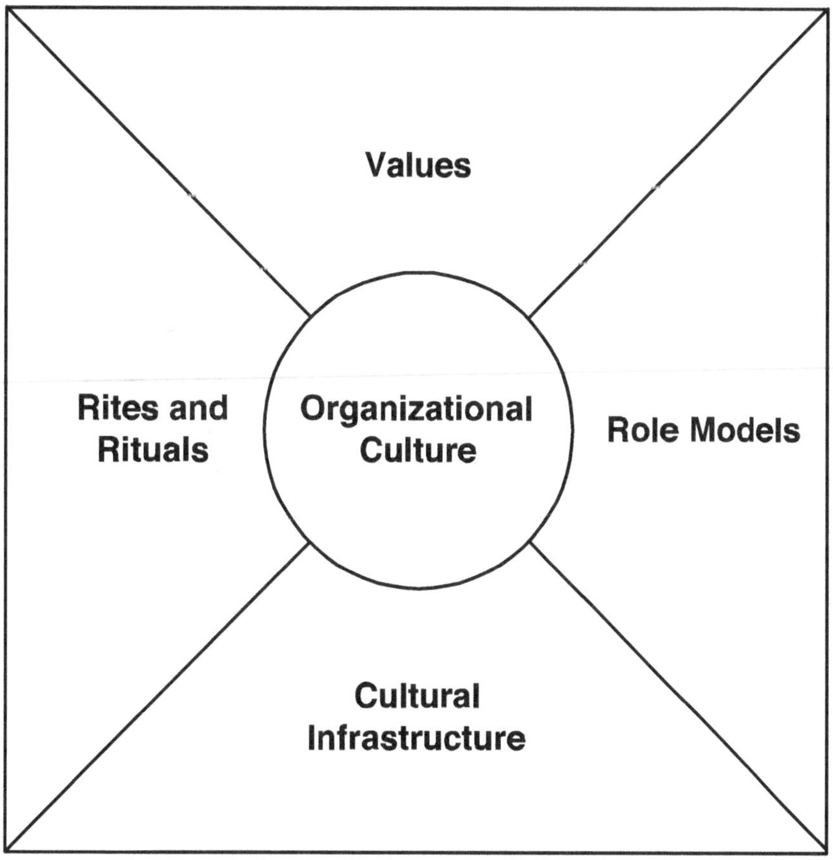

Figure 2-5 The Four Elements of Culture

VALUES
define how we act

ROLE MODELS
provide examples of desired performance

RITUALS AND RITES
what we do and how we reinforce it

CULTURAL INFRASTRUCTURE
behind the scene activities that support the values

Exercise 2-3 The Definition of Culture Dissected

The left column breaks the definition of culture down into its component parts. What do these various parts mean to you? Provide examples.

Definition	Group or Individual Meaning
A pattern of shared basic assumptions	
that the group learned as it solved its problems of external adaptation and internal integration	
that has worked well enough to be considered valid	
and, therefore, to be taught to new members as the correct way to perceive, think, and feel in relation to those problems.	

Exercise 2-4 Recognizing the Parts of the Culture

Chapter 2 introduced the four elements of culture. In this exercise, identify which of these four elements of culture match each scenario. Enter OV for organizational values, RM for role models, RR for rites and rituals, and CI for cultural infrastructure. It is possible that more than one element may apply to a scenario. Use the one that is the closest match.

Scenario	Part
The reliability manager continuously promotes reliability repairs; as a result, people have begun to make them.	
When the quick fix is completed and production is re-established, the work crews get praised for their efforts.	
The work process includes planning, work scheduling, and field execution.	
At every meeting, one of the senior managers always relates stories about how things used to be when he started.	
"Rapid repair is how we work around here. When things break, it is our job to get them back into service. We don't have time for detailed analysis."	
The field execution leader is always on top of the most recent equipment failure and mobilizes the workforce to make the fix. Lately, others have started to work closely with production so that when something fails, they can quickly respond.	
The company provides high quality service awards to employees for every five years of service.	
All new employees are assigned a more senior manager who has the responsibility to guide their development for the first five years of their employment.	
Everything that is said at the morning maintenance meeting is communicated to the manager before the superintendent has a chance to report it.	
Root cause failure analysis is conducted on every major equipment failure before the repair strategy is devised.	

Figure 2-6 How the Organization Acts Toward Change

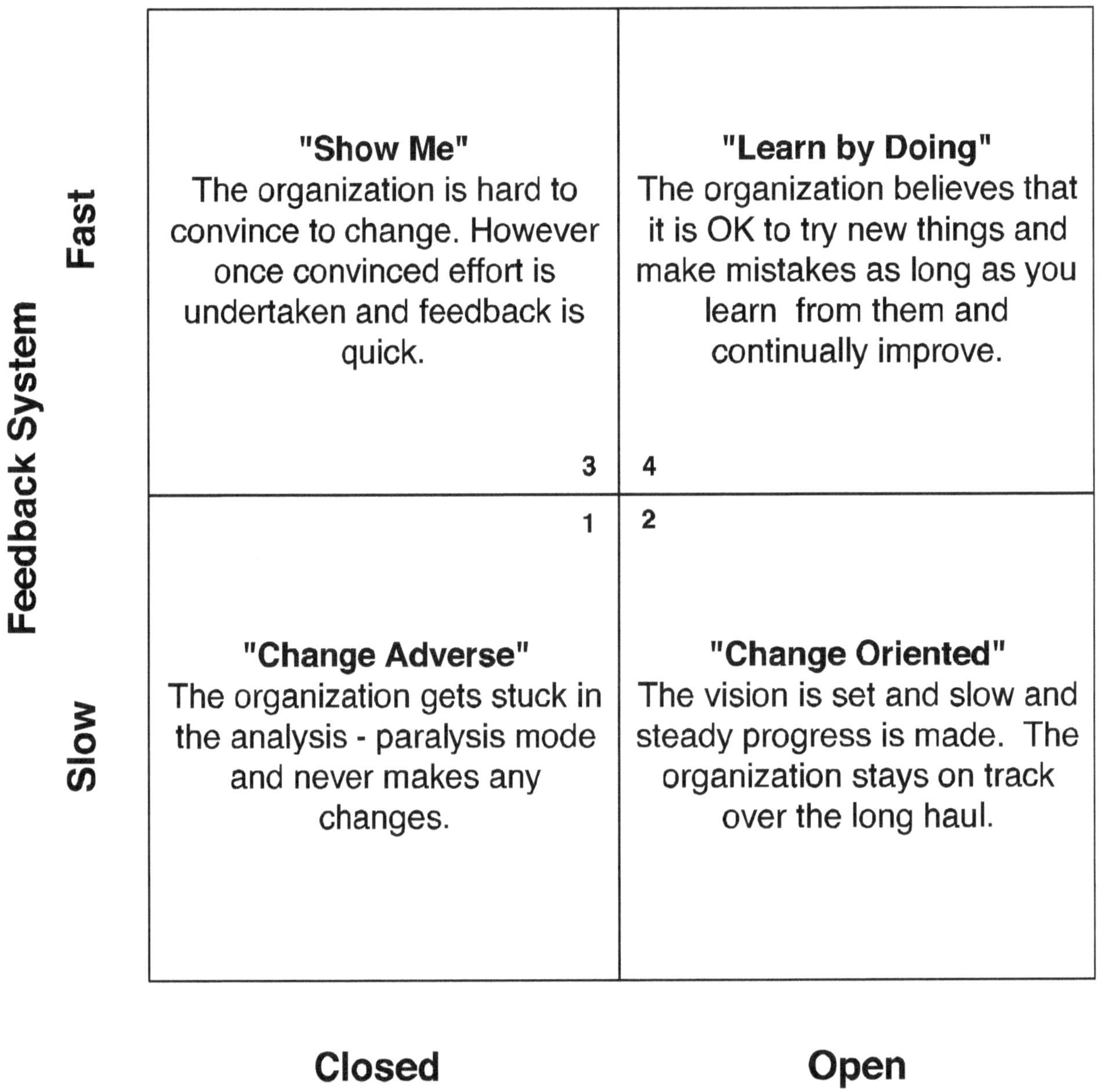

Exercise 2-5 Organizational Culture Types

Figure 2-6 shows various cultural types based on how they respond to change. Answer the following ten questions to determine which type best describes you. For each question, 1 = strongly disagree, 2= disagree, 3= neutral, 4 = agree and 5 = strongly agree. Enter your selection in the unshaded block. When you have answered all questions, total the scores for each column in the boxes at the bottom of the page.

Questions	Your Answer			
It is hard to initiate change in the organization, but once started, we follow through to completion.				
No matter what I propose, my management never has enough information to make a decision.				
We have been successful in making change over the long term based on our unaltered vision.				
The organization believes it is a good idea to try new things.				
After a lot of hard work, we establish the vision and then work on making progress towards its attainment.				
We try new things and learn from both the positive and negative outcomes.				
We never seem to be able to make any change; we are always analyzing the data to support it.				
We provide ourselves with rapid feedback on our change efforts so that we can adjust as needed.				
Column Totals				

Exercise Analysis

In the row below, copy your column totals from the previous page. Then write them again in the unshaded block for each column.

Column Totals				
Cultural Type	↓	↓	↓	↓
Change Adverse The organization gets stuck in the analysis – paralysis mode and never makes any changes.				
Change Oriented The vision is set; slow and steady progress is made. The organization stays on track over the long haul.				
Show Me The organization is hard to convince to change. However, once convinced, effort is undertaken and feedback is quick.				
Learn by Doing The organization believes that it is OK to try new things and make mistakes as long as you learn from them and continually improve.				

Translate your scores to the chart on the next page. A sample is provided.

Exercise 2-5 continued

Using your scores from the previous page, chart your cultural type. The top chart below provides a sample.

Sample:

Your Scores:

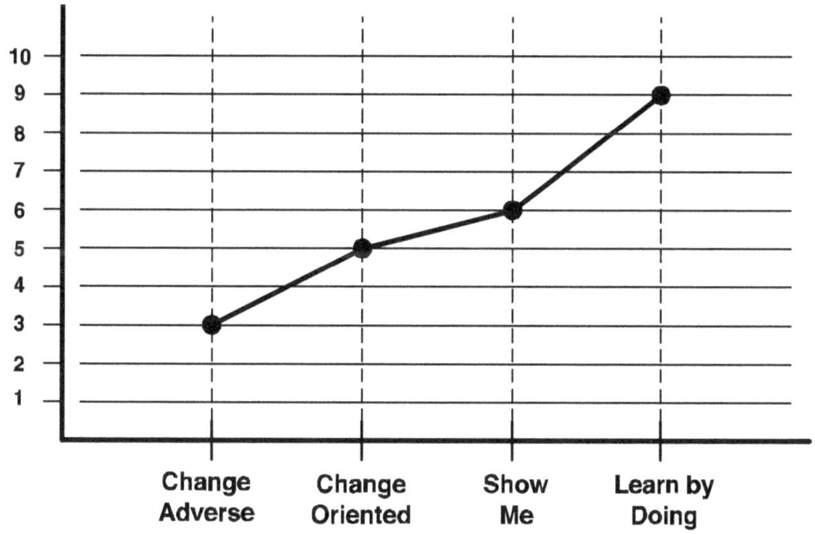

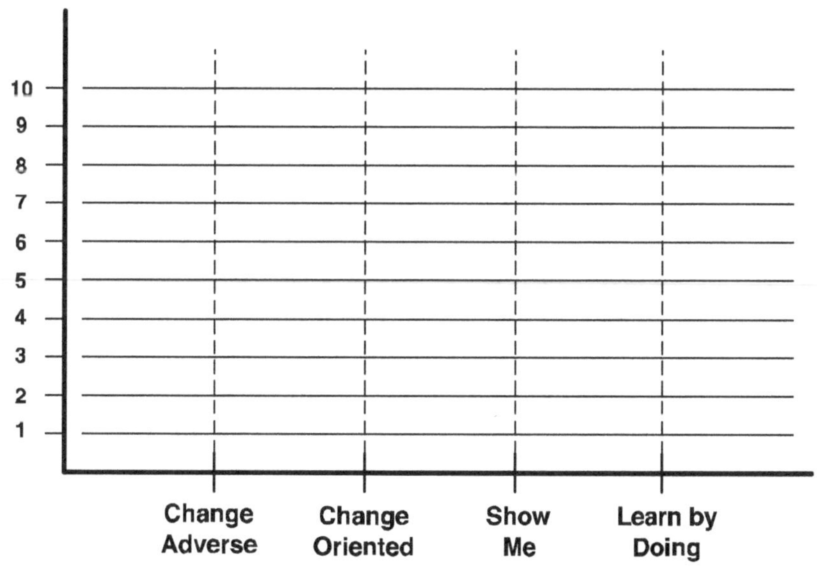

The results indicate the type of organizational culture at work. Some results will have scores predominantly in one area; others will show more of a mixture. There is no right mix; however, the results of the exercise should help you identify the type of culture in which you work.

Exercise 2-6 ATPCo and Their Culture - Part 1

Suggestion: Before completing this exercise, read Appendix 1; it provides a detailed description of ATPCo.

Mike Kane, the plant manager, has worked in the eastern plant since it went into production ten years ago. He has always believed in and followed the vision established by company president and friend, Alex Peters. This vision-operate the facilities in a manner that optimizes production and is capable of effectively and efficiently meeting the demand of the market-is very clear in its intent for Mike. He interprets the vision by running the equipment hard to meet demand; then, if it breaks, he fixed it as quickly as possible in order to return it to service.

For ten years ago, this strategy has worked very well. With limited competition, no new entries into the eastern market, and an unlimited supply of raw material, the "break it – fix it" maintenance strategy has always enabled the plant to meet the demand. Additionally, the plant is relatively new. Therefore, it has few if any reliability problems that require anything other than repair-in-kind maintenance.

Mike and his staff promote his work ethics by

- Assigning mechanics and engineers to the production line instead of keeping all of these skills within maintenance.
- Hiring people who have demonstrated their ability to respond quickly to the demands of the production department.
- Measuring the time between equipment failure and its return to service as a prime performance metric.
- Providing rapid repair training for the maintenance group.
- Removing from the organization those individuals who promote other alternative courses of action such as preventive or predictive maintenance .
- Personally mentoring new hires to the management team through weekly meetings and coaching session.
- Personally overseeing major repair efforts.
- Second guessing the decisions of his staff.
- Communicating his expectations with his favorite statement “my way or the highway.”

Exercise 2-6 Part 1 continued

Answer the following questions regarding ATPCo's work culture.

1. What are the shared basic assumptions that drive this culture?

2. How were these basic assumptions learned by the organization?

3. What were the external and internal forces or events that integrated these concepts into the culture?

4. Why did the culture that emerged work well?

5. How are people taught the culture?

6. How was the culture reinforced by Kane and others on his staff?

Exercise 2-6 ATPCo and Their Culture Part 2

As you read in the appendix, ATPCo has a problem due to the emergence of a new product called Epsilon coupled with their inability to reliably meet market demand for all of their products.

Discussion:

1. What basic assumptions are no longer valid and why?

2. What should be the new assumptions for the eastern plant?

3. How would you suggest that the eastern plant unlearn their old ways and adopt a new culture with the new assumptions?

Exercise 2-7 Your Organization's Culture

Answer the following questions about your plant's culture.

1. What are the shared basic assumptions about reliability and maintenance within your plant? These assumptions need to be supported with factual data.

2. How did your organization learn and create these shared assumptions? What were the problems it solved to develop them?

3. What were the external and internal events that validated the basic assumptions?

4. How did the organization define things that "worked well" so that the basic assumptions were validated?

5. How are the basic assumptions taught to new members?

6. How are non-accepted actions corrected (those that deviate from the basic assumptions)?

Chapter 3

Vision and the Goal Achievement Model

Purpose

This chapter introduces two key concepts that are a necessary part of a successful change process: the vision and the Goal Achievement Model. The vision paints a picture of the desired future state. The Goal Achievement Model lays out the roadmap to get you there.

What Is Included

Figure 3-1	The Change Hurdle Diagram
Figure 3-2	Vision Defined
Exercise 3-1	Understanding the Definition of Vision
Exercise 3-2	What Is Missing?
Exercise 3-3	Correcting the Vision
Exercise 3-4	Do You Walk the Talk?
Figure 3-3	The Workings of the Goal Achievement Model
Figure 3-4	The Relationship Between the Elements of the Goal Achievement Model and Strategic and Tactical Work
Figure 3-5	Components of the Goal Achievement Model
Figure 3-6A	A Sample of the Goal Achievement Model
Exercise 3-5	Building ATPCo's Goal Achievement Model
Exercise 3-6	Building Your Own Goal Achievement Model

Figure 3-1 The Change Hurdle Diagram

This chapter focuses on hurdles #2 and #3.

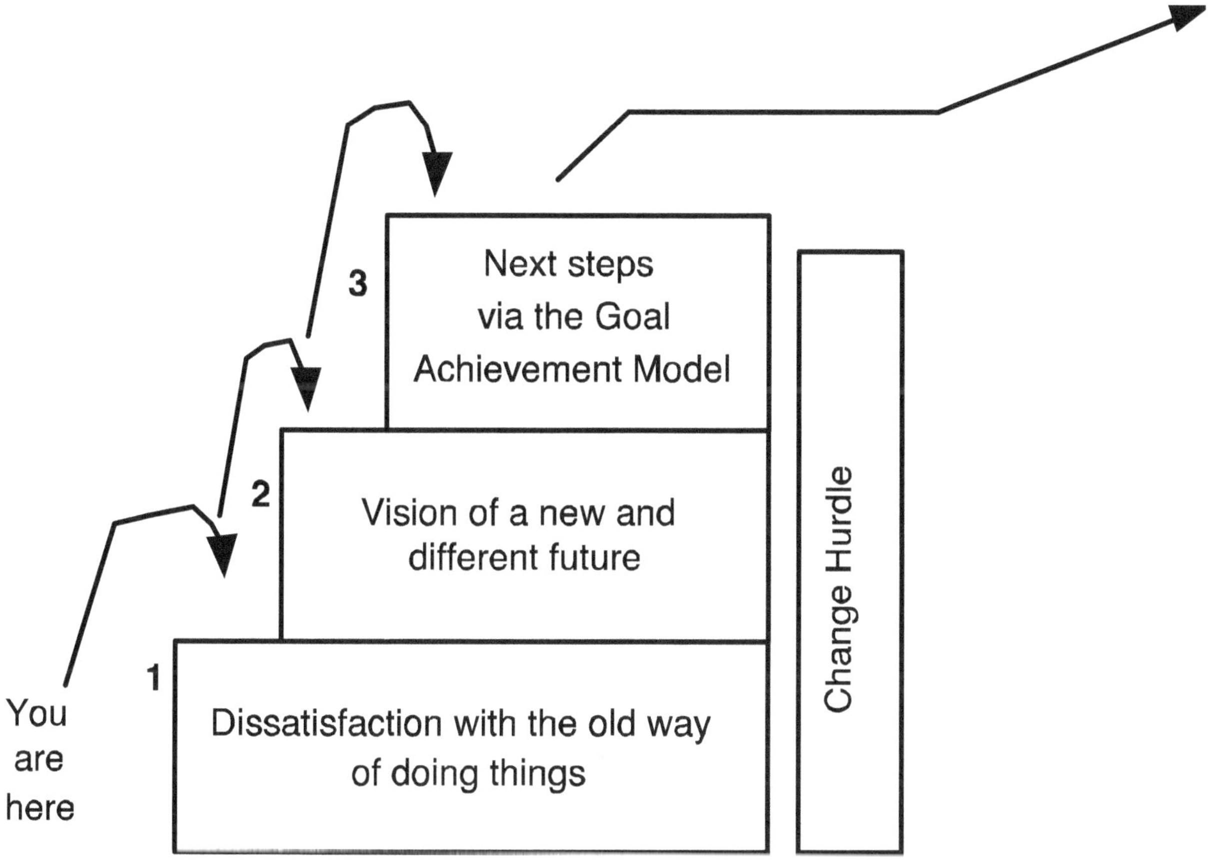

Figure 3-2 Vision Defined

Vision is an
idealized picture of a future state,
one that is
integrated into the organization's culture.
It provides a
stretch, yet it is achievable
over an extended time period with
a great deal of work and collective focus
by the entire organization. Because it
continually evolves,
it is never fully achieved.

Exercise 3-1 Understanding the Definition of Vision

Describe what each of the individual components of the definition of vision (Figure 3-2) means to you. This exercise can be completed individually, but working in a group setting is preferable. The idea is to better understand the meaning which, in turn, will either help you create a good vision or improve upon the one you already have in place.

Vision	Meaning
Idealized picture of a future state	
Integrated into the organization's culture	
A stretch – yet achievable	
Extended time (how much is needed to achieve the vision?)	
A great deal of work (what does this mean to you?)	
Collective focus	
Continually evolves	

Exercise 3-2 What Is Missing?

Now that you have discussed the definition of vision, determine what is missing from the following vision statements.

Vision	What Is Missing?
Create a plant PM program to improve our equipment reliability.	
Treat the reliability of our equipment the same as we treat the safety of our employees.	
Get back to basics – do them correctly, then reliability will follow.	
Operate our plant with a focus on safety and the reliability of the equipment.	
Optimize production to maximize profit.	

Exercise 3-3 Correcting the Vision

For this exercise, use the vision statements from Exercise 3-2 and the deficiencies that you have identified. Correct, modify, or even rewrite them to be more in line with the definition of vision.

Vision	The New Vision
Create a plant PM program to improve our equipment reliability.	
Treat the reliability of our equipment the same as we treat the safety of our employees.	
Get back to basics – do them correctly, then reliability will follow.	
Operate our plant with a focus on safety and the reliability of the equipment.	
Optimize production to maximize profit.	

Exercise 3-4 Do You Walk the Talk?

Groups have difficulty implementing strategic efforts without ongoing leadership. As one of your organization's leaders, you need to provide visible, continuous commitment to accomplishing the vision. This is referred to as "walk the talk."

This exercise helps you measure how well you walk the talk. It can also be given to your managers or subordinates so that they can see how well they are showing visible commitment to achieving the vision.

Mark the appropriate block **Yes** or **No** for each of the following questions:

Question – As a manager, I....	**Yes**	**No**
1. Make certain that the vision exists and is upgraded as the group learns.		
2. Seek opportunity to communicate the vision at all levels.		
3. Attempt to show how my employees' work relates to the vision.		
4. Review progress towards accomplishing the vision on a frequent basis.		
5. Set clear expectations for work performance aligned with the vision.		
6. Work to sell the vision and its value to others wherever and whenever possible.		
7. Attend meetings to show visible support when initiatives that address the vision are being implemented.		
8. Identify and provide training to support the vision and related activities.		
9. Make decisions that show support of the vision rather than decisions that do not show support.		
10. Take positive, corrective action to correct deficiencies.		
Total Yes and No columns		

Evaluation: If you have 9 or 10 Yes answers, then you walk the talk. A score of 7 or 8 Yes answers indicates that your approach is not bad, but some improvement is needed. If you score less than 7 Yes answers, you should re-evaluate your approach.

Figure 3-3 The Workings of the Goal Achievement Model

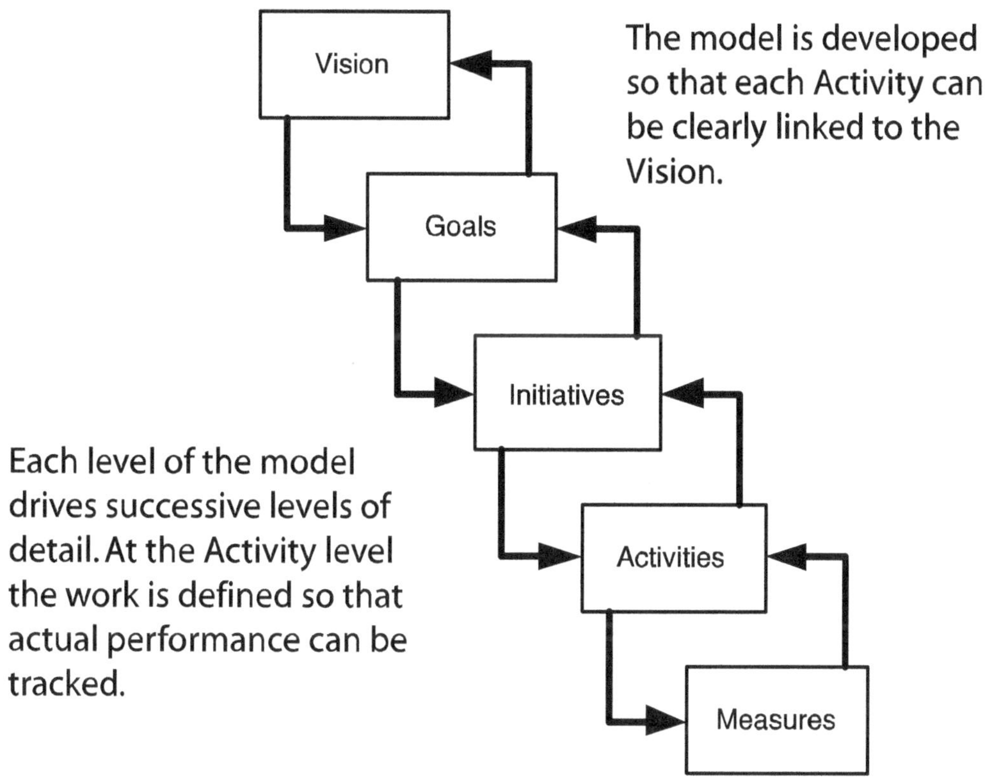

Figure 3-4 The Relationship Between the Elements of the Goal Achievement Model and Types of Work (Strategic or Tactical)

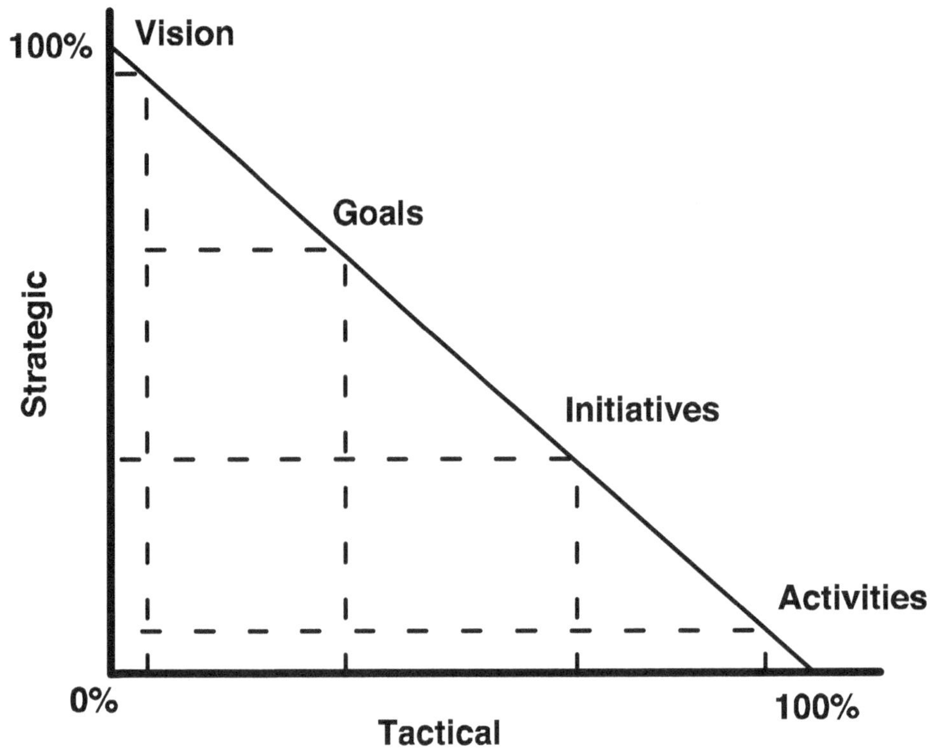

Figure 3-5 Components of the Goal Achievement Model

Vision – A high-level strategic statement of what you wish to accomplish for the organization, department or work team.

Goals – Major elements of the work strategy that when taken together accomplish the vision. A missing goal will invariably increase the likelihood of failure.

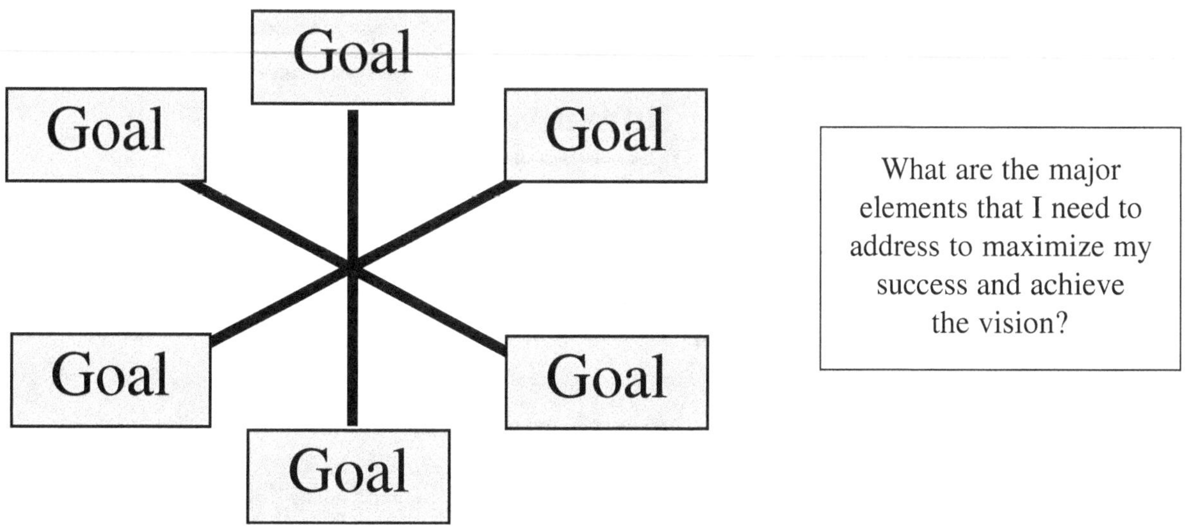

Figure 3-5 Components of the Goal Achievement Model (continued)

Initiatives – Within each goal, what specific components are required so that I can initiate action in the desired direction to accomplish the goal?

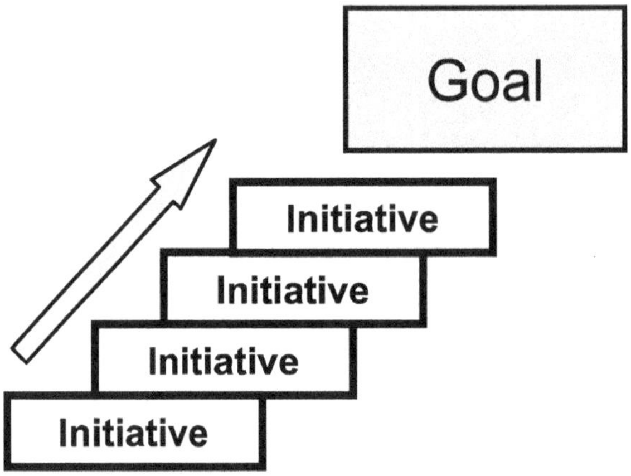

Activities – What very specific steps do I need to take to have the initiative succeed? This is the "how do we do it" part.

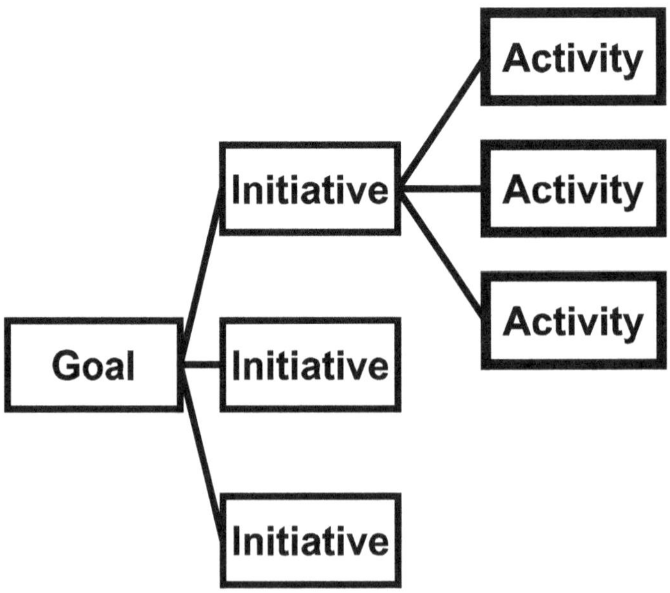

Figure 3-5 Components of the Goal Achievement Model (continued)

Measures – What are the very specific measurements that I need to show me the progress of each activity?

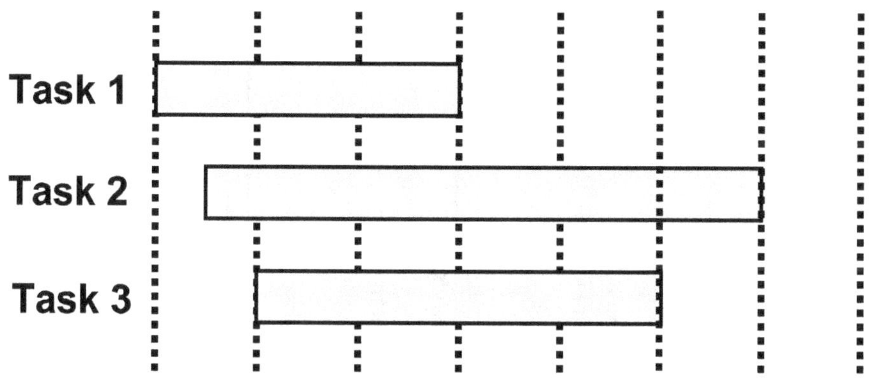

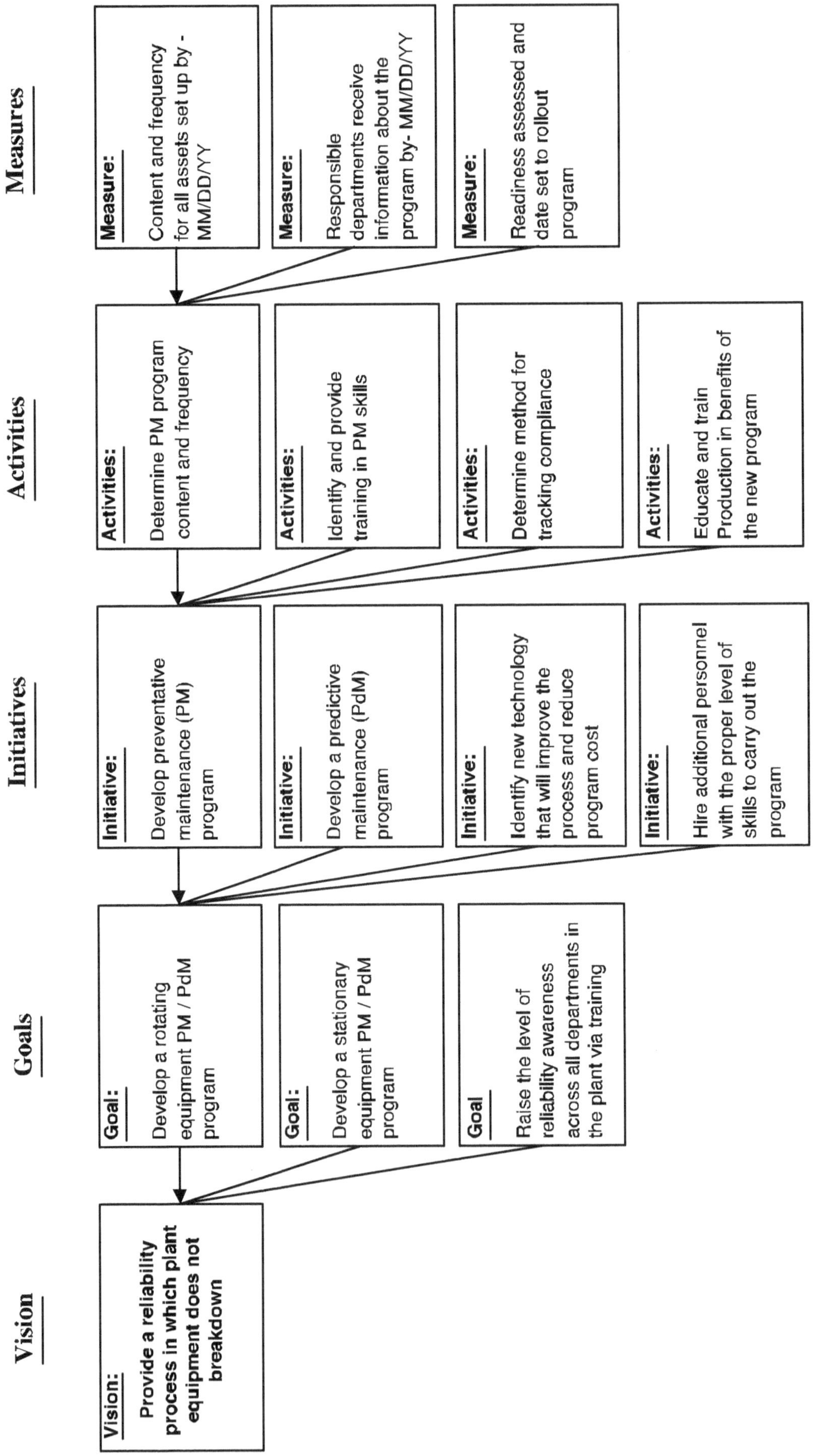

Figure 3-6 A Sample of the Goal Achievement Model

Exercise 3-5 Building ATPCo's Goal Achievement Model

ATPCo has developed a new vision based on the philosophy of its new plant manager. That vision is: *Operate the facility in a manner that optimizes production and uses the equipment in a reliable manner in order to meet the demands of the eastern market.* This new vision is a major change for those at ATPCo. The new manager is reliability focused and completely opposed to the "break it, fix it" mentality of his predecessor.

Furthermore, he completely understands the concept of the Goal Achievement Model from its extensive use at his former place of employment. It would be easy for him to give his staff a detailed, fully-developed model, but he wants them to learn how to develop and apply it themselves. As a result, he provides them with the following information.

Vision - *Operate the facility in a manner that optimizes production and uses the equipment in a reliable manner in order to meet the demands of the eastern market.*

Goals:

1. Create a reliability-focused work process.
2. Improve the structure of the organization to support equipment reliability.
3. Teach the organization reliability concepts; use these concepts as a fundamental part of the work.
4. Build a communication process that reinforces reliability-based performance within the plant.

Exercise 3-5 Develop the initiatives, activities, and measures. Use the form provided as a work sheet.

Exercise 3-5 Worksheet

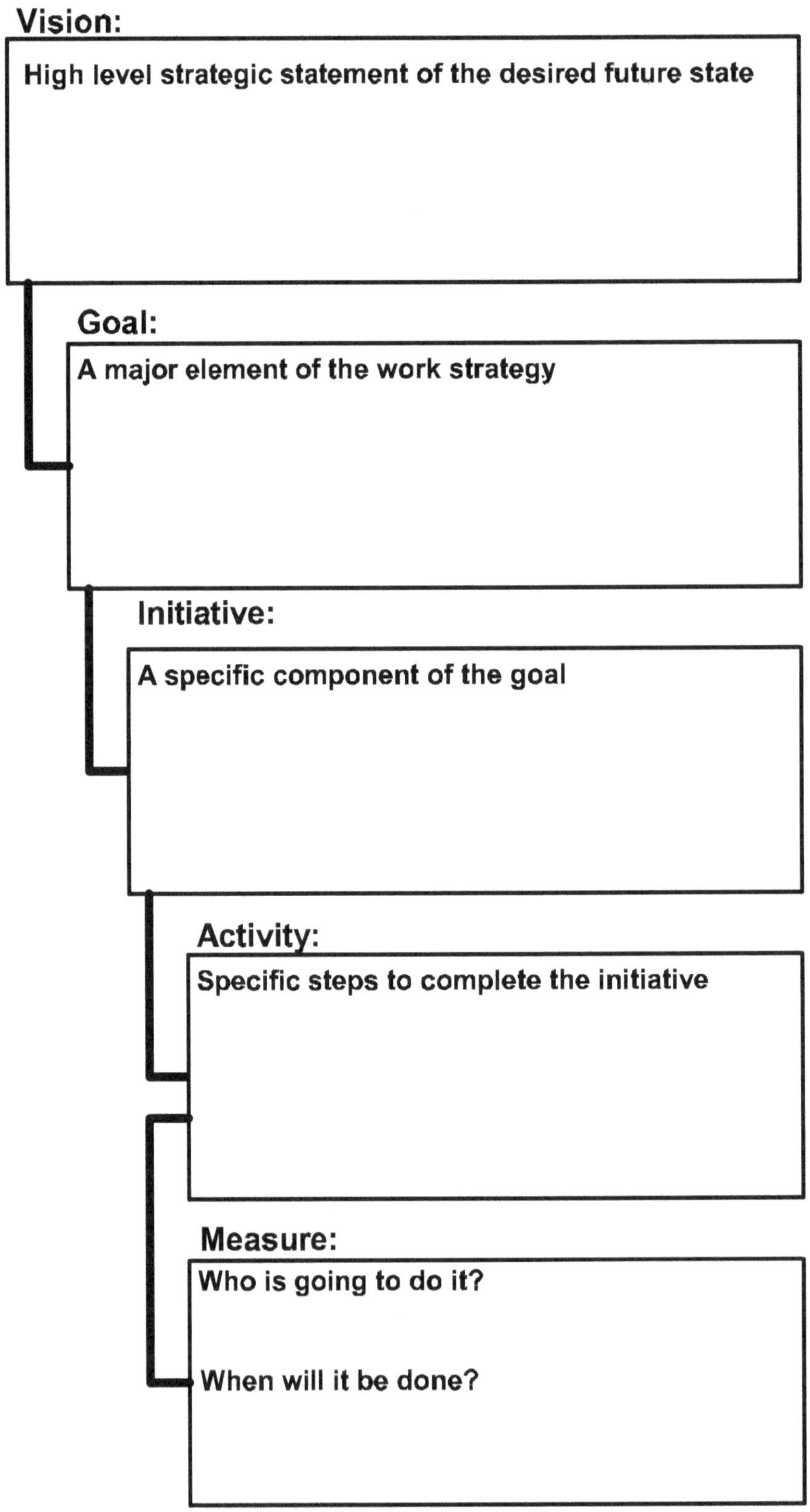

Exercise 3-6 Building Your Own Goal Achievement Model

For the Goal Achievement Model to work successfully, all of its parts need to be completed. Then everyone in the organization can clearly see the vision as well as understand how their individual and group work fits into the scheme of things. The next diagram provides the format for this exercise. Use it to build a Goal Achievement Model in Excel. State your vision, develop the supporting goals, list the initiatives supporting these goals, identify the activities supporting each initiative, and list the measures tracking the activities. By completing this diagram for all goals, initiatives, and activities, you will clearly see how the Goal Achievement Model fits into the overall process. Parts that are missing will be evident; then corrective action can be initiated.

Notes

VISION:

GOALS	INITIATIVES	ACTIVITIES	MEASURES

Chapter 4

Organizational Values

Purpose

This chapter introduces the first of the four elements of organizational culture organizational values and looks at how this element applies to your company, your department, your work team, and you as a member.

What Is Included

Figure 4-1 Organizational Values Defined

Exercise 4-1 What Are Your Reliability / Maintenance Organizational Values?

Exercise 4-2 What Are Your Reliability / Maintenance Values?

Figure 4-2 Steps for Changing the Organizational Values

Figure 4-3 Vision vs. Values

Exercise 4-3 Values Gone Astray

Exercise 4-4 Expressed vs. Actual Values

Figure 4-4 Resistance

Exercise 4-5 Resistance to Organizational Values

Exercise 4-6 Organizational Values and the Goal Achievement Model

Exercise 4-7 The Organizational Values Web

Figure 4-5 A Full-Size Organizational Web Diagram

Figure 4-1 Organizational Values Defined

A company's basic, collectively understood, universally applied, and wholly accepted set of beliefs about how to behave within the context of the business. They also describe what achieving success feels like. These values are internalized by everyone in the company and, therefore, are the standard for expected behavior.

Exercise 4-1

What Are Your Reliability / Maintenance Organizational Values?

Read the ten difference scenarios. Then identify what you and your organization would really do in this situation. Also identify your response as R (reactive) or P (proactive) maintenance.

Scenario	What Would You Really Do?	R= Reactive P=Proactive
An important pump fails; the back-up is available. You are ready to fix it, but are not sure of the true failure mechanism.		
The preventive maintenance program is falling behind schedule and your manager wants to pull the crew for routine work.		
A new reliability technology is introduced in your industry. You believe it will benefityour plant, but no money is in the budget.		
One of your key supervisors is scheduled for training. The program has been on her schedule for the last six months. You are having plant problems and are short-handed.		
You want to introduce a detailed planning and scheduling process. The heart of the program involves scheduling planned work for the coming week and not deviating from the schedule. Production complains that there are too many urgent jobs needing to get done now.		

Scenario	What Would You Really Do?	R= Reactive P=Proactive
You have tried to implement an equipment reliability program, but have not been successful. The main reason is that the workforce is always focused on today's emergency. Production tells you how happy they are about your rapid response to their problems.		
During a production line outage, you discover a problem that you believe requires immediate attention. The production manger tells you that he thinks it will last until the next outage and that maintenance, if done now, will cost the company a great deal of money in lost profit.		
You see some mechanics circumventing a procedure. You know it will get the job done sooner–exactly what you are under extreme pressure to do.		
You have to choose one of two foremen for promotion. They are equal in every way except their approach to the work. One likes to consider the options and make reliability-based repairs. The other is known for the "quick fix." Production likes the latter of the two and has been very vocal.		
Your mechanics discover missing replacement parts for an important repair. Do you re-use the old ones, even if they are questionable, or delay production and wait for the new parts to be delivered?		

Exercise 4-2 What Are Your Reliability / Maintenance Values?

In Exercise 4-1, you stated what you and your organization would do in various reliability / maintenance scenarios. Select several of your answers from that exercise, then respond to the following questions.

1. In your plant, what are the basic beliefs about reliability and maintenance? These beliefs will define your organizational values and current culture.

2. How are these values communicated so that people instinctively know how to act in various situations?

3. Are the values universally accepted? If not, why not? Furthermore, if not, why do people still act according to those values?

4. How are new employees trained to behave in the manner dictated by the values?

5. What does success feel like in this work environment?

Figure 4-2 Steps for Changing the Organizational Values

After completing Exercises 4-1 and 4-2, you may clearly see the need to change the organizational values that are governing your behavior and actions. Easily said, but not easily done. The five steps below provide a good way to start. My books Successfully Managing Change in Organizations: A Users Guide and Improving Maintenance and Reliability Through Cultural Change guide you through the process

STEP 1 Develop a clear and concise vision of the future.

STEP 2 Communicate the vision continuously to the organization.

STEP 3 Build goals, initiatives, and activities in support of the vision.

STEP 4 Hold people accountable to the work they are doing in Step 3.

STEP 5 Reward those in alignment and punish those who are not. Maintain zero tolerance for deviation.

Figure 4-3 Vision vs. Values

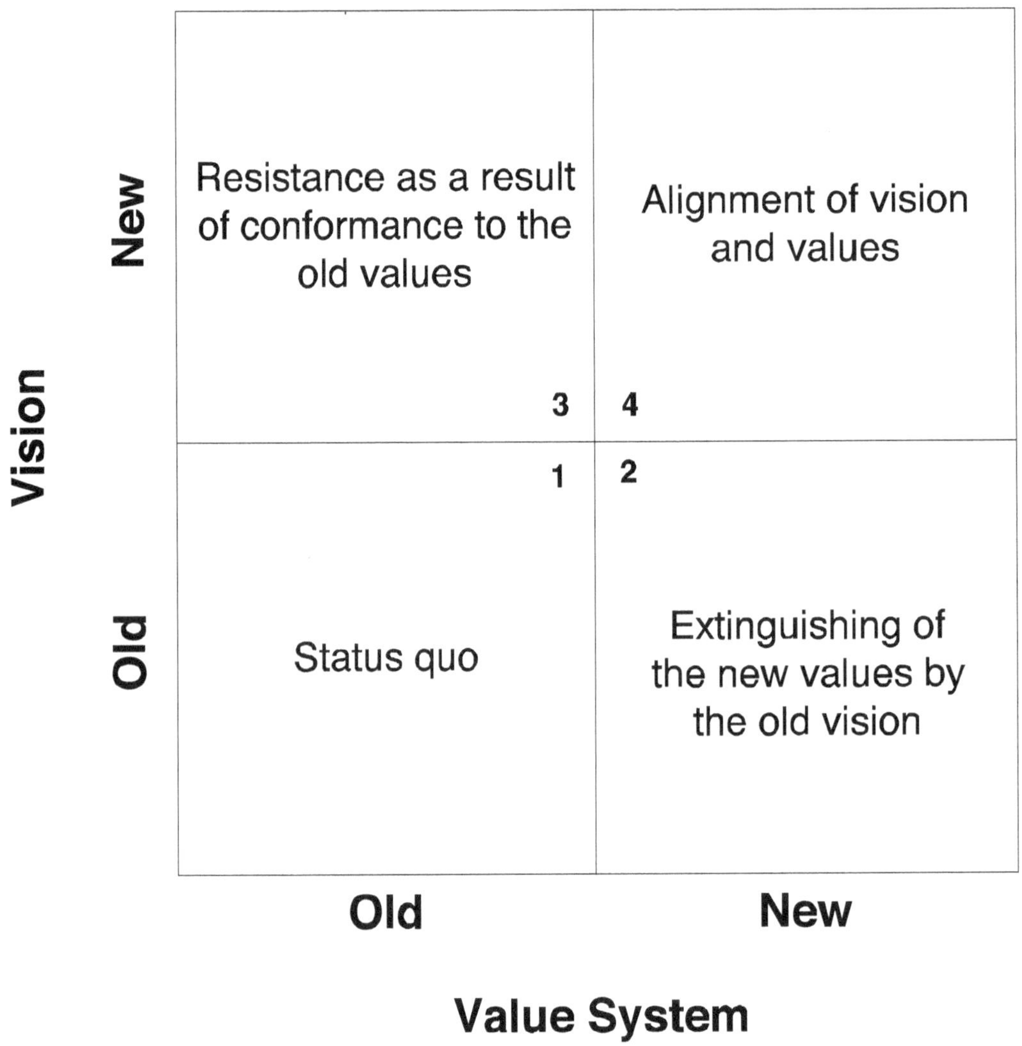

Exercise 4-3 Values Gone Astray

Often an organization's values can get off track. This can happen for several reasons: 1) the values are no longer relevant; 2) they are inconsistently applied; 3) they simply are not followed and everyone knows it; and 4) the values are valid, but people resist them. Recognizing these issues when they occur is important. In your opinion, which of these four potential reasons for getting off course apply to each of the following scenarios?

Scenario	Reason #
The company states that it wants all repairs to have a reliability component. However, most repairs are defined simply as "replace what has broken and get back into production as soon as possible."	
The company states that planning and scheduling of the work is a key success factor. However, rapid response to production's needs (real or unreal) is how the plant works and how people are rewarded.	
The company tells everyone that safety is the most important part of any job, but people still violate the safety rules and are not reprimanded.	
Preventive maintenance plans must be completed before these crews can do other non-PM work. Yet the crews in Area 1 are always pulled away from the PM prematurely.	
Management has stated that planning will be a part of every job, yet the work plans are never given to the crews. Instead, they often are found in the trash can in the foremen's office.	
The plant manager has gone on record stating that reliability will be factored into every job. You know from personal experience that the reliability engineers are not always brought onto the job in time to effect repairs of this type.	
The production manager likes the fact that the maintenance organization reports directly to him so that he can direct what gets done. He holds to this belief even though his competition has set up separate maintenance groups to focus on reliability.	
The reliability manager has stated "we will optimize the use of reliability technology." Nevertheless, you know that the plant still does things the way they did ten years ago and the reliability tools go unused.	

Exercise 4-4 Expressed vs. Actual Values (Part 1)

A wide discrepancy often exists between the written and the actual behavioral values within a company. When both are in alignment, a company or plant can be considered healthy from the standpoint of organizational values and work culture. When they are not, people are often torn between them. This exercise asks you to evaluate your written values against what actually takes place on the shop floor. Examples are shown on page 48 of the text. In the left column, state the written values in use at your plant. Then in the right column, describe what actually happens-the actual value system at work. These actual values may or may not match the plant's expressed values.

Written Values	What Actually Happens

Exercise 4-4 Expressed vs. Actual Values (Part 2)

Based on what you wrote in Part 1 of this exercise, rate your written values vs. those in practice.

Rate Yourself: Read each statement from Part 1, then circle the numbered rating that applies. You are evaluating how much the written and actual value system is in alignment.

Written vs. Actual	Rating
The written values and those actually in place are identical	5
The written values and those actually in place are the same most of the time.	4
The written values and those actually in place are the same only some of the time.	3
The written values and those actually in place are often not the same.	2
The written values and those actually in place are almost never the same.	1

Complete the exercise: Based on your answers thus far, respond to the following two questions.

1. Why do you believe that the written values and the actual way that they are used are different?

2. What can you do to bring them into alignment?

Figure 4-4 Resistance

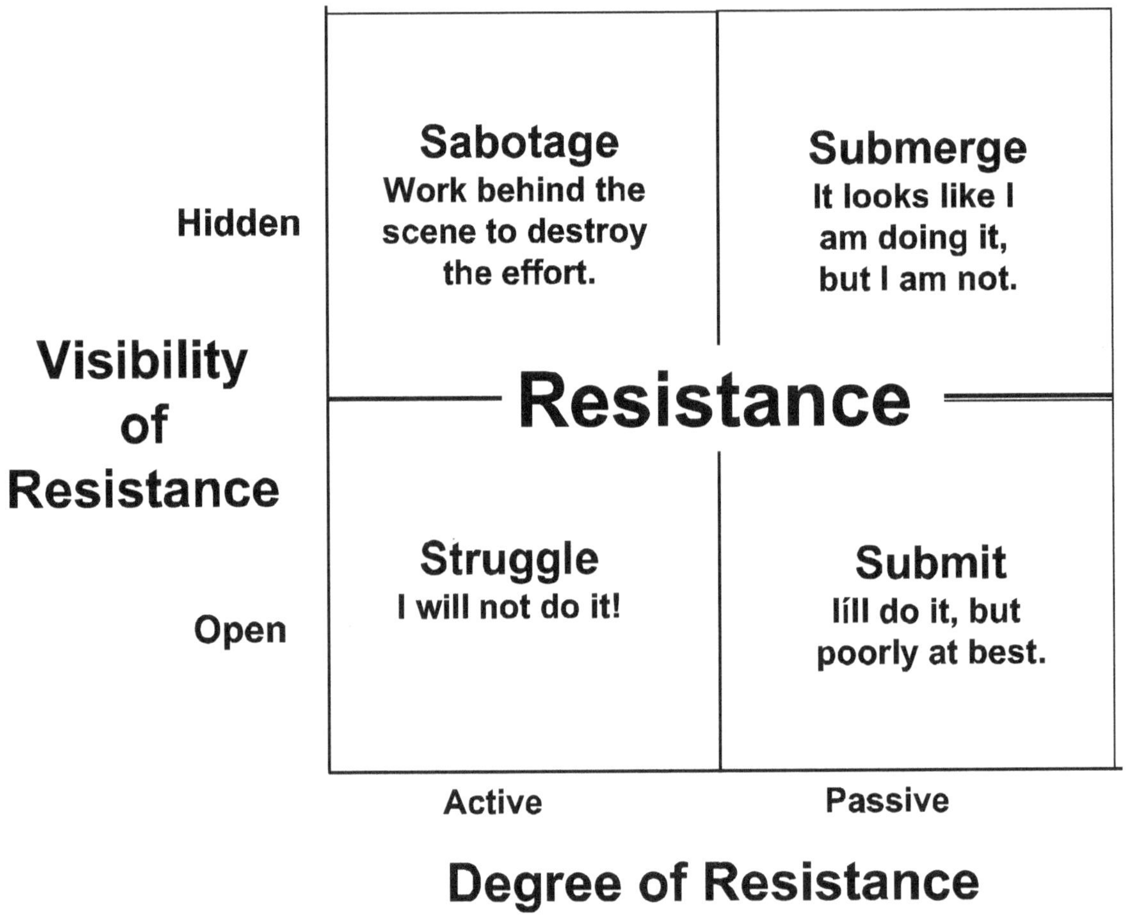

Exercise 4-5 Resistance to Organizational Values

Under new leadership, ATPCo has worked diligently to shift from a reactive to a new reliability-focused set of organizational values. However, things don't seem to be working out as planned. In this instance, as in your own plant, it is important to recognize the various and often subtle forms of resistance at work. Using Figure 4-4, identify the type of resistance for each of the following scenarios.

Stated Value	What Is Actually Happening	Type
Planning and scheduling will be an integral part of every job.	Planning and scheduling takes place, but the plans are poor and the schedule is never followed.	
No repair will be made without consideration for improving its reliability.	Reliability engineers are purposefully not told about work in the field until it is too late	
Training is important. Everyone will spend 10 percent of their time per year in programs to help them develop their skills.	People are scheduled, but they either don't show up or leave early.	
Preventive and predictive maintenance is a key aspect of how we work. The crews with these assignments will not be pulled to other jobs.	Crews are almost always pulled by the foremen themselves in order to respond to production's problem-of-the-day.	
Any break in the weekly work schedule is considered an emergency and will be tracked as such.	People agree, but often do not follow the guideline. Deviation is frequent. Often the foremen will not even report the schedule changes.	

Stated Value	What Is Actually Happening	Type
We will purchase and utilize new and improved tools to support the re liability efforts	The tools are purchased, but mysteriously break or disappear.	
Our computer-based maintenance system is important and we will optimize its use.	The equipment is not used. People openly refuse to stop using their antiquated paper systems.	
Repair history and equipment documentation is important to our reliability efforts and will be maintained.	Most people refuse to document their work, stating that there is no value. Those who do don't do very good job.	
Production will meet weekly with maintenance to establish next week's schedule.	The two groups meet, but key players don't show up for the meeting.	
Work will not be started until all material is on the job to prevent interruptions or delays.	Crews start work knowing that the parts are not available. This has caused production issues for the plant.	

Exercise 4-6 Organizational Values and the Goal Achievement Model

The Goal Achievement Model introduced in Chapter 3 of *Improving Maintenance and Reliability Through Cultural Change* is an important tool to mange the change process. It can also be used to examine in detail the culture needed to support change.

Part 1 of Exercise 4-6 examines a portion of the Goal Achievement Model under the leadership of Mike Kane. It asks you to determine the organizational values at work in the Mike Kane model. Part 2 provides you with the reliability-based model of the new management team, and Part 3 asks you to discuss how you could make the shift from one to the other.

Part 1

Under the leadership of Mike Kane, ATPCo was entirely focused on uptime of the production lines regardless, of the longer-term reliability impact on the plant. A small portion of the Goal Achievement Model he used to drive his vision run it hard, fix it fast follows.

Vision: Optimize production to meet the demand of the market place.

Goals (one of many)	Initiatives	Activities
Keep the production lines running. Address line shutdowns quickly in order to return lines to service	Create a rapid-repair work process so that line shutdowns are instantly addressed by maintenance.	Activities in all cases support execution of the initiatives.
	Have people working in maintenance who are experienced in rapid repair strategies.	
	Train mechanics in rapid repair strategies focused on the "quick fix" in order to return the line to service.	
	Assign the best mechanics as " line mechanics" assigned to production so that they are always available to handle problems as they occur.	

Questions

1. What are Mike Kane's basic beliefs, his organizational values? In addition to the goals above, see Appendix 1.

2. How do you think that these values would be communicated and taught to the plant population?

3. Within the context of these organizational values, what do you think Mike believes is a successful maintenance effort?

Exercise 4-6, Part 2

Given the introduction of the new product Epsilon and the declining reliability of the eastern plant, the president of ATPCo, Allen Peters, has replaced Mike with another manager. Allen recognized that Mike's way of working was hurting the company and that a new reliability-focused approach needed to be taken. The new manager has developed the following set of goals.

<table>
<tr><td colspan="3">Vision: Run the plant reliably and supply the market without interruption, without the need to import products from other plants, and without the ability to add equipment.</td></tr>
<tr><th>Goals
(one of many)</th><th>Initiatives</th><th>Activities</th></tr>
<tr><td rowspan="3">Keep the production lines running through the use of reliability techniques that eliminate equipment failure.</td><td>Institute a focused planning and scheduling process to optimize the maintenanc effort.</td><td rowspan="3">Activities that are developedwill directly support the completion of the initiatives.</td></tr>
<tr><td>Have people working in maint enance who are experienced in reliability strategies.</td></tr>
<tr><td>Train mechanics in reliability techniques focused on failure elimination.</td></tr>
</table>

Questions

1. What are the basic beliefs of the new plant manager–his organizational values?

2. How do you think that these values will be communicated and taught to the plant population?

3. Within the context of these organizational values, describe a successful maintenance effort.

Exercise 4-6, Part 3

Mike Kane ruled by fear and by practicing his "my way or the highway" philosophy. This philosophy was furthered by his hiring of people with the same beliefs. If you were the new manager, what would you do to change the organizational values to those stated in Part 2? List the activities and measures you would implement to support the new set of initiatives.

Initiative	Activities	Activity Measures

Exercise 4-7 The Organizational Values Web

Chapter 17 explains the Web of Cultural Change; it indicates that for each of the eight elements of change (the radial spokes of the diagram), there are sixteen questions to be answered to create the score for that element. The first four questions within each of the elements are those that address organizational values. Therefore, by answering these four specific questions for each of the eight elements of change. you can build a web diagram specific to the organizational values.

This exercise is divided into three parts. The first part asks you to answer the organizational values questions for each of the eight elements of change. In Part 2, you use your scores from Part 1 to create your organizational values web of change. Then in Part 3, you analyze the web diagram by selecting those elements that received a low score and asking yourself (or your work group) why those scores were as low as they appeared. Chapter 18 in the text can help you accomplish this analysis.

Exercise 4-7 The Organizational Values Web, Part 1

Leadership **Total Score** ________

1 Leadership has established a reliability-focused value system.

A. Strongly agree	5
B. Agree	4
C. Neutral	3
D. Disagree	2
E. Strongly disagree	1

2 Leadership has communicated this value system throughout the plant.

A. Strongly agree	5
B. Agree	4
C. Neutral	3
D. Disagree	2
E. Strongly disagree	1

3 The value system at the site has been audited to assure that it is understood at all levels of the organization.

A. Strongly agree	5
B. Agree	4
C. Neutral	3
D. Disagree	2
E. Strongly disagree	1

4 The site leadership demonstrates through their behavior and actions that they believe andare following the established values.

A. Strongly agree	5
B. Agree	4
C. Neutral	3
D. Disagree	2
E. Strongly disagree	1

Work Process **Total Score** ________

1 The work process has been designed to support the organizational values.

A. Strongly agree	5
B. Agree	4
C. Neutral	3
D. Disagree	2
E. Strongly disagree	1

2 The work process is periodically audited to assure alignment, detect issues, and address corrective action.

A. Strongly agree	5
B. Agree	4
C. Neutral	3
D. Disagree	2
E. Strongly disagree	1

3 The work process includes the ability to accurately capture information about the plant assets in such a way that the information can be utilized in support of the process and related values.

A. Strongly agree	5
B. Agree	4
C. Neutral	3
D. Disagree	2
E. Strongly disagree	1

4 The work process has clearly delineated how the work flow and information flow models align with each other.

A. Strongly agree	5
B. Agree	4
C. Neutral	3
D. Disagree	2
E. Strongly disagree	1

Structure **Total Score** __________

1. The structure has been designed to reflect the organizational values.

A. Strongly agree 5

B. Agree 4

C. Neutral 3

D. Disagree 2

E. Strongly disagree 1

2 Reliability and maintenance are understood to be a responsibility of multiple organizations and not a business function. This shared responsibility is clearly reflected in the values.

A. Strongly agree 5

B. Agree 4

C. Neutral 3

D. Disagree 2

E. Strongly disagree 1

3 Line and staff work issues are adequately handled by the structure. The structure was designed to make this possible.

A. Strongly agree 5

B. Agree 4

C. Neutral 3

D. Disagree 2

E. Strongly disagree 1

4. When the structure was redesigned for the new process, the effort involved more than simply rearranging the organizational boxes. Serious thought was put into the effort.

A. Strongly agree 5

B. Agree 4

C. Neutral 3

D. Disagree 2

E. Strongly disagree 1

Group Learning **Total Score** __________

1. Type 1 (single loop) learning is used to support current goals.

A. Strongly agree	5
B. Agree	4
C. Neutral	3
D. Disagree	2
E. Strongly disagree	1

2. The company or site is open to Type 2 (double loop) learning even though employing it may identify that its goals are incorrect.

A. Strongly agree	5
B. Agree	4
C. Neutral	3
D. Disagree	2
E. Strongly disagree	1

3. Gaps identified in either of the two types of learning processes are handled by developing corrective action items and follow up to completion.

A. Strongly agree	5
B. Agree	4
C. Neutral	3
D. Disagree	2
E. Strongly disagree	1

4. One of the organizational values is the need to learn. The process is in place to assure that learning is continuously taking place.

A. Strongly agree	5
B. Agree	4
C. Neutral	3
D. Disagree	2
E. Strongly disagree	1

Technology **Total Score** __________

1. The technology has functionality that supports the organizational values.

A. Strongly agree	5
B. Agree	4
C. Neutral	3
D. Disagree	2
E. Strongly disagree	1

2. When decisions need to be made, the technology readily provides the needed data to the users.

A. Strongly agree	5
B. Agree	4
C. Neutral	3
D. Disagree	2
E. Strongly disagree	1

3. There is a close tie between the continuous improvement plan for the organization and the needed technology to support it.

A. Strongly agree	5
B. Agree	4
C. Neutral	3
D. Disagree	2
E. Strongly disagree	1

4. When the organizational values are upgraded (or changed), the technology, a critical component, is also upgraded.

A. Strongly agree	5
B. Agree	4
C. Neutral	3
D. Disagree	2
E. Strongly disagree	1

Communication **Total Score** ________

1. The organizational values are widely communicated.

A. Strongly agree	5
B. Agree	4
C. Neutral	3
D. Disagree	2
E. Strongly disagree	1

2. There are on-going processes in place to assure that the organizational values are understood.

A. Strongly agree	5
B. Agree	4
C. Neutral	3
D. Disagree	2
E. Strongly disagree	1

3. Where questions exist about the organizational values and their interpretation, feedback is provided for clarification.

A. Strongly agree	5
B. Agree	4
C. Neutral	3
D. Disagree	2
E. Strongly disagree	1

4. There is a continuous process in place to keep the workforce informed about the organizational values and how they should be applied in the day-to-day work.

A. Strongly agree	5
B. Agree	4
C. Neutral	3
D. Disagree	2
E. Strongly disagree	1

Interrelationships **Total Score** ________

1. Organizational values are widely shared by the leadership team to promote buy-in and positive interrelationships.

A. Strongly agree	5
B. Agree	4
C. Neutral	3
D. Disagree	2
E. Strongly disagree	1

2. People at all levels work well together and are open to change.

A. Strongly agree	5
B. Agree	4
C. Neutral	3
D. Disagree	2
E. Strongly disagree	1

3. There is a single set of organizational values that is applied across all functions. There are no functional silos where the values apply differently, dependent on your function.

A. Strongly agree	5
B. Agree	4
C. Neutral	3
D. Disagree	2
E. Strongly disagree	1

4. If audited, the site would show clear evidence that one set of values existed and was, in fact, being applied.

A. Strongly agree	5
B. Agree	4
C. Neutral	3
D. Disagree	2
E. Strongly disagree	1

Rewards **Total Score** __________

1 The organizational values include a reward process that supports its success.

A. Strongly agree	5
B. Agree	4
C. Neutral	3
D. Disagree	2
E. Strongly disagree	1

2. The organization clearly understands the reward process as well as how it is linked to the organizational values.

A. Strongly agree	5
B. Agree	4
C. Neutral	3
D. Disagree	2
E. Strongly disagree	1

3. The Goal Achievement Model (or a similar process) is used to promote the values through the vision-goals-initiative-activity process.

A. Strongly agree	5
B. Agree	4
C. Neutral	3
D. Disagree	2
E. Strongly disagree	1

4. The system provides negative or neutral rewards for courses of action that do not support the value system being utilized.

A. Strongly agree	5
B. Agree	4
C. Neutral	3
D. Disagree	2
E. Strongly disagree	1

Exercise 4-7 The Organizational Values Web, Part 2

Enter the score from each of the elements into the table below.

Elements	Max Score	Baseline Score
Leadership	20	
Work Process	20	
Structure	20	
Group Learning	20	
Technology	20	
Communication	20	
Interrelationships	20	
Rewards	20	

Chart your scores on the web diagram to create your organizational value web.

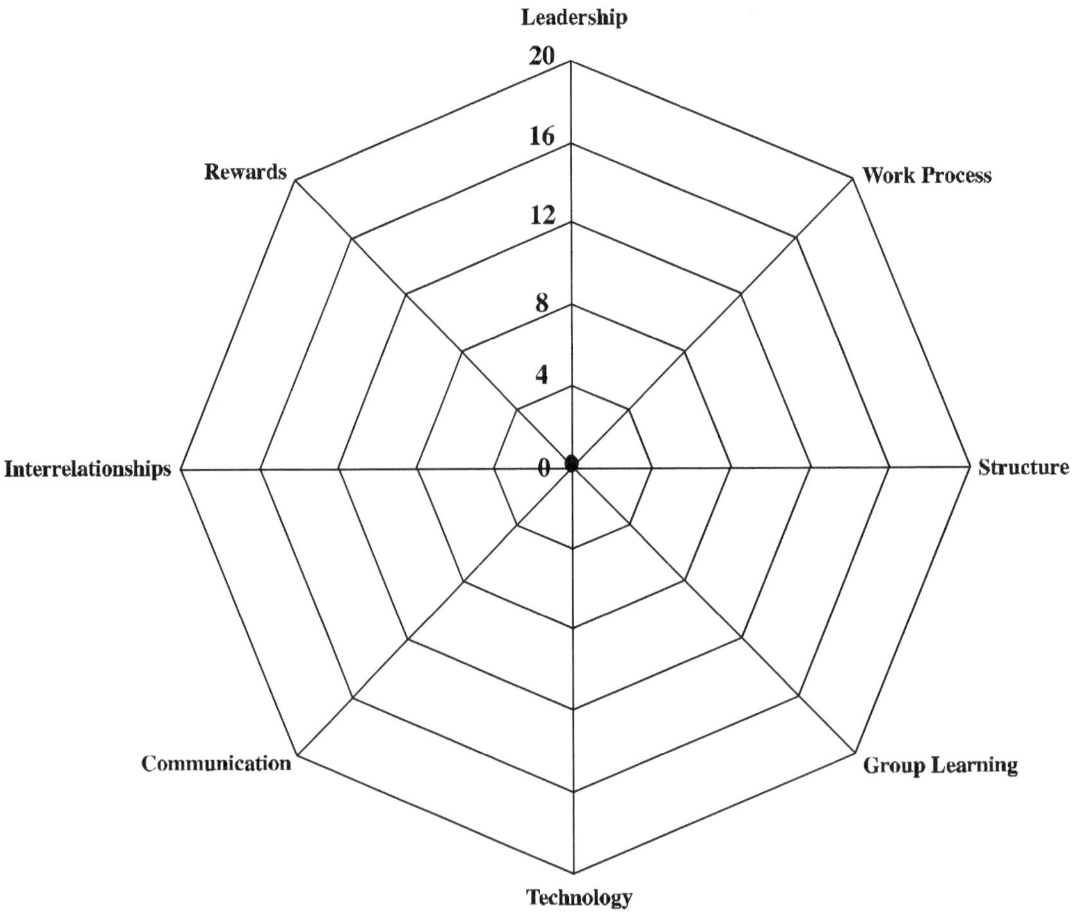

Exercise 4-7 The Organizational Values Web, Part 3

Explain why each of the eight elements of change related to organizational values has scored the way that it did. Describe what needs to be done to improve those scores that fall below 12.

Elements	Problem	Corrective Action
Leadership Score =		
Work Process Score =		
Structure Score =		
Group Learning Score =		
Technology Score =		
Communication Score =		
Interrelationships Score =		
Rewards Score =		

Figure 4-5 A Full-Size Organizational Web Diagram

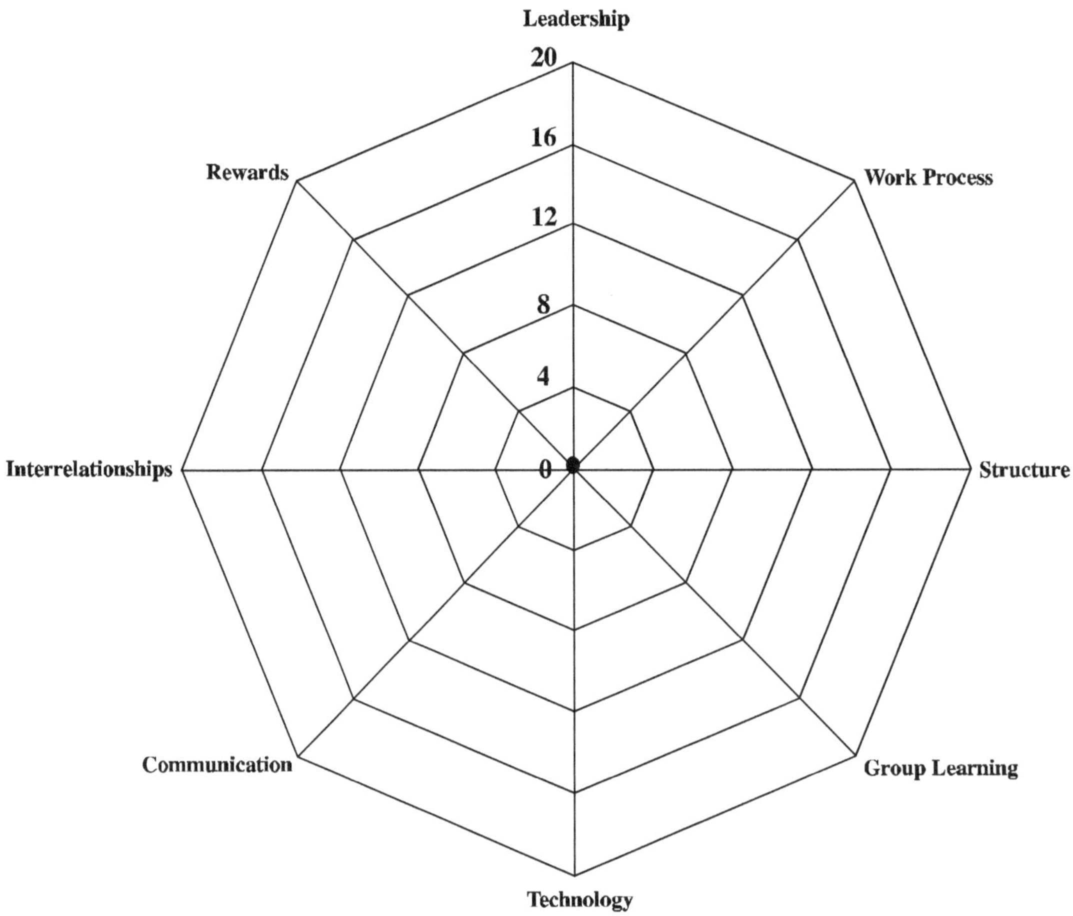

Role Models

Chapter 5

Purpose

This chapter introduces the concept of cultural role models and looks at how they fit within the world of organizational culture and change management.

What Is Included

Figure 5-1	Role Models Defined
Exercise 5-1	Who Are Your Role Models?
Figure 5-2	Role Models: Types of Work vs. Work Focus
Figure 5-3	Strategic vs. Tactical Role Models
Exercise 5-2	Role Model Types
Figure 5-4	Role Models in a Reactive Culture
Figure 5-5	Role Models in a Proactive Culture
Exercise 5-3	How Role Models Fit the Culture
Figure 5-6	The Role Model's Journey from Reactive to Proactive
Exercise 5-4	The Role Model's Journey
Exercise 5-5	The Role Model Web
Figure 5-7	A Full-Size Organizational Web Diagram

Figure 5-1 Role Models Defined

Role models are people within organizations who exhibit traits that appeal to us and which we can apply to how we conduct our business. These role models are usually at or near the top of the organization; they have been successful within the organizational culture. They demonstrate a successful behavior style within the business culture, one that we feel comfortable adopting as our own.

Exercise 5-1 Who Are Your Role Models?

Based on the definition in Figure 5-1, identify who your role models are and why you chose them. Being able to identify people within your company who you emulate will help you clarify the definition of a role model.

Name
Position
Have your role models been successful within the organization? If so, why?
What traits do they exhibit that appeal to you?
What behavior style do they use as they perform their job?
Are you comfortable adopting their style?
How do you apply this style to your work?

Figure 5-2 Role Models: Type of Work vs. Work Focus

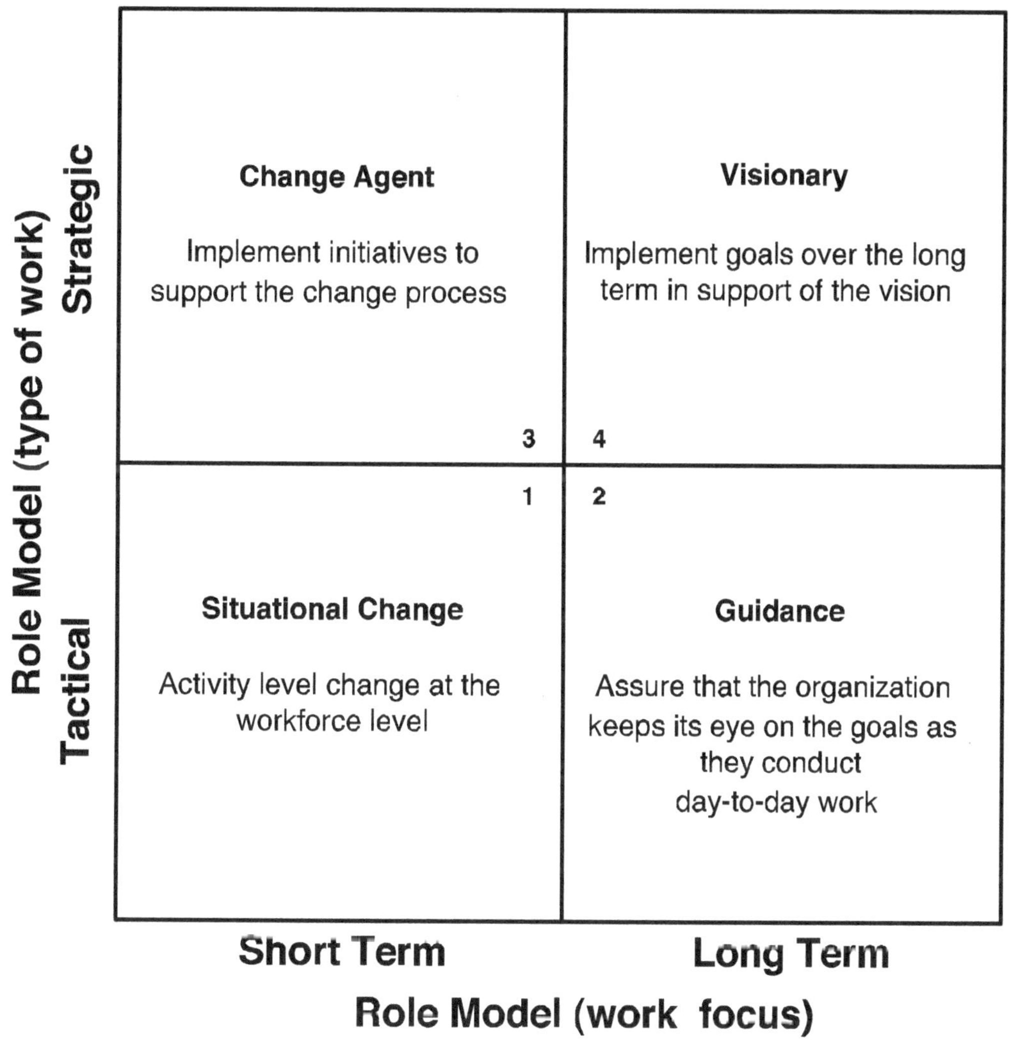

Figure 5-3 Strategic vs. Tactical Role Models

This figure shows how the work performed by role models fits within the Goal Achievement Model.

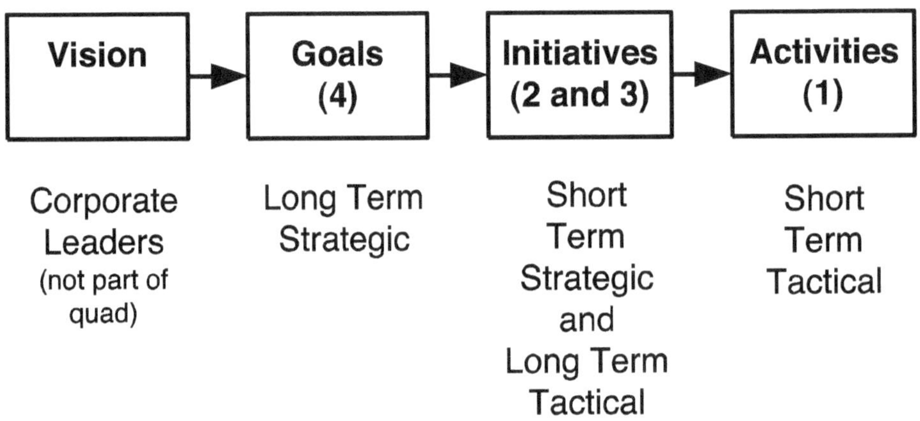

Exercise 5-2 Role Model Types

Figure 5-2 compares role models' work focus (short or long term) and the types of work they conduct (tactical or strategic). In Exercise 5-1, you identified one or more individuals in your firm who were your role models. Now conduct Exercise 5-2 to determine the type of role model.

Part 1

If the role models fit the description provided, write Yes in the unshaded box. If they do not fit the description, write No in the unshaded box.

When you have completed the form, count the yes answers for each column and write the totals in the bottom row. The numbers in the column headers match the four sections of Figure 5-2; they will give you some insight into the type of role model you have identified.

Description	1	2	3	4
Their focus is short term, dealing primarily with day-to-day work efforts.				
Their focus is long term, dealing with goals and initiatives that foster change over time .				
They implement initiatives in the day-to-day work that support change.				
Their focus is both long term and strategic. There is little day-to-day work involvement.				
Their focus is tactical (day-to-day), addressing and solving plant problems.				
They work on day-to-day initiatives, but never lose site of the long-term goals.				
Their primary focus is strategic and designed to promote improvement to attain the company vision.				
They support change in the day-to-day work with minimal involvement in longer-term initiatives.				
Count the yes answers and enter the totals here:				

Exercise 5-2 Role Model Types, Part 2

Now complete the same exercise for yourself to determine your style. Once again, count the yes answers and show the totals at the bottom of the form.

Description	1	2	3	4
My focus is short term, dealing primarily with day-to-day work efforts.				
My focus is long term, dealing with goals and initiatives that foster change over time.				
I implement initiatives in the day-to-day work that support change.				
My focus is both long term and strategic. There is little day-to-day work involvement.				
My focus is tactical (day-to-day), addressing and solving plant problems.				
I work on day-to-day initiatives, but never lose site of the long-term goals.				
My focus is strategic and designed to promote improvement to attain the company vision.				
I support change in the day-to-day work with minimal involvement in longer-term initiatives.				
Count the yes answers and enter the totals here:				

Exercise 5-2 Role Model Types, Part 3

An important point regarding role models is that you can emulate their behaviors and work practices more if the two of you are doing the same type of work. For instance, if you and your role models are change agents (working in the short term, yet focused on strategic initiatives), it would be far easier for you to emulate their behavior than if they were a visionary (quad 4) and you worked in the arena of situational change (quad 1). In part of this exercise, you compare your position in the quad diagram with that of your role models.

Quad	Role Model Score	Your Score	Difference
Quad 1 Situational Change			
Quad 2 Guidance			
Quad 3 Change Agent			
Quad 4 Visionary			

Where the differences in the scores are small, you can reasonably assume that you can adopt many aspects of the role models' behavior. However, be aware that if the differences are great, you may need to chose different role models for your short-term development.

Figure 5-4 Role Models in a Reactive Culture

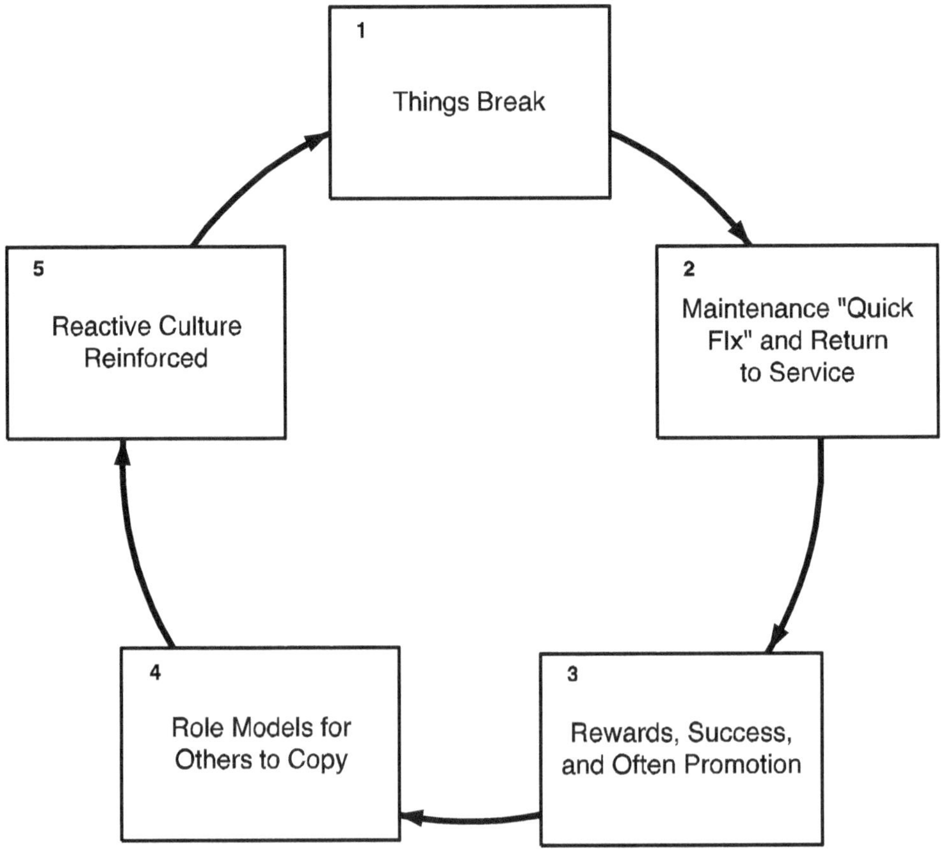

Figure 5-5 Role Models in a Proactive Culture

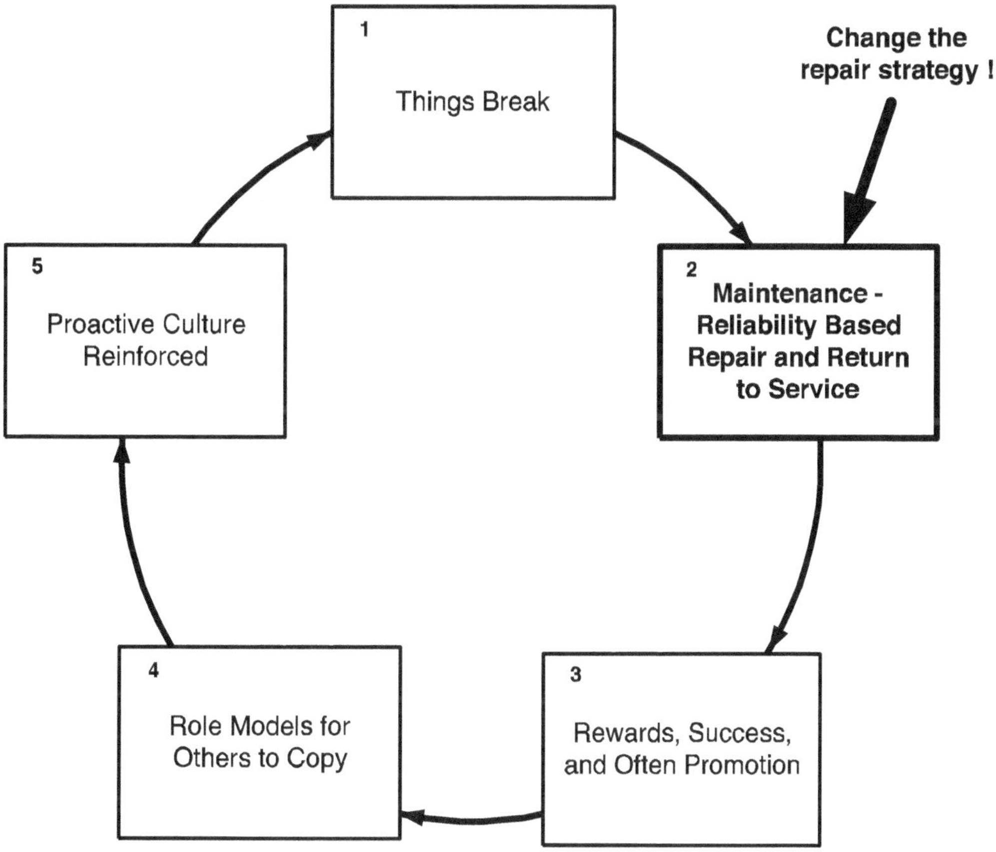

Exercise 5-3 How Role Models Fit the Culture

It is important to understand how role models fit into their culture and support its continuation. It is also important to understand what you need to do, related to role models, if you wish to alter the culture. Using Figures 5-4 and 5-5, answer the following questions about reactive role models who need to adapt to a proactive, reliability-focused work environment.

1. In Figure 5-4, how would the role model support the existing reactive work culture? Examples where you observed a role model in this situation would be valuable to support your reasoning.

2. In Figure 5-5, explain the problems the existing reactive role models will have with a new maintenance culture, as indicated in block #2 of the figure. What are the things that you would have to do as a manager to bring about a different behavior from these role models?

Figure 5-6 The Role Model's Journey from Reactive to Proactive

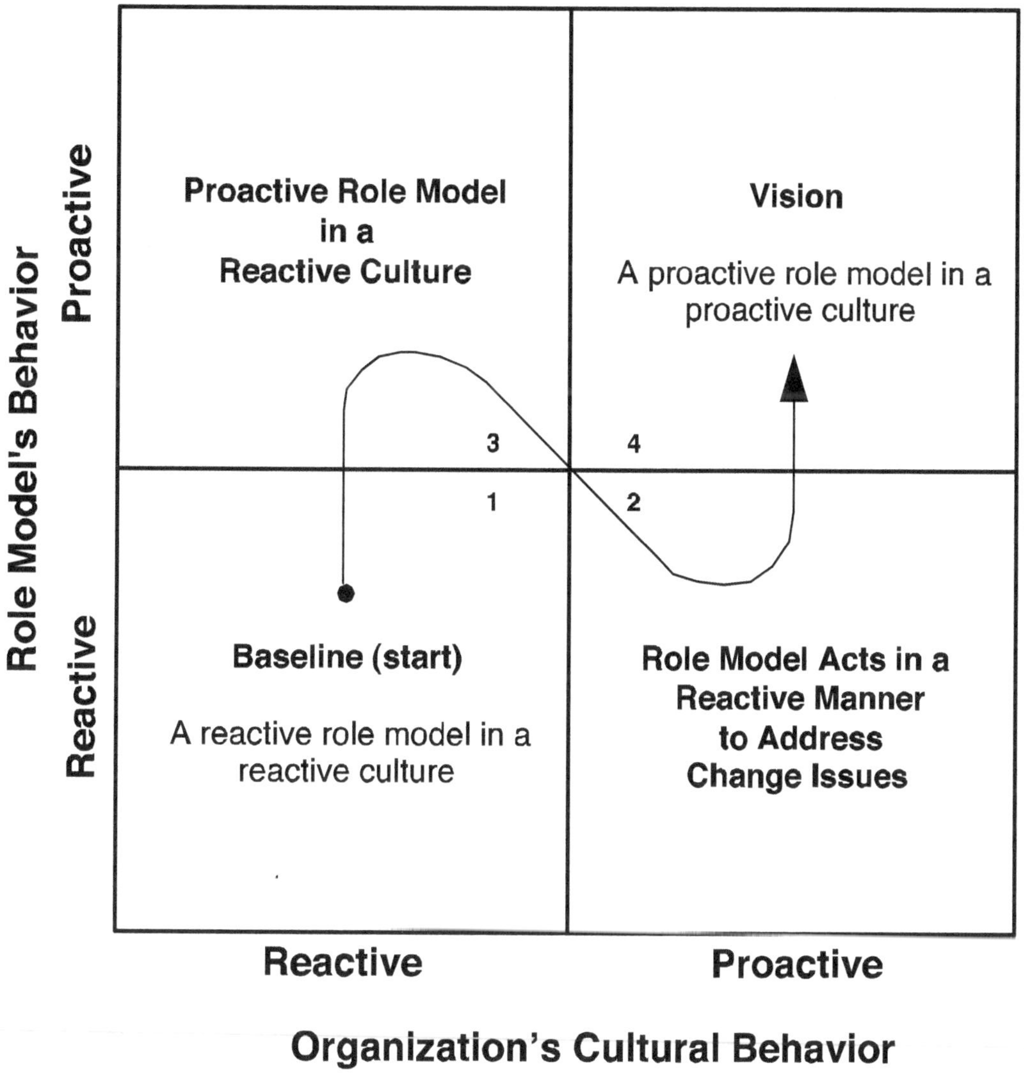

Exercise 5-4 The Role Model's Journey

As the new manager at ATPCo's eastern plant, you have been charged by Allen Peters, ATPCo's president, to implement new work processes designed to improve plant reliability. Certainly the changes you need to make run counter to the culture left behind by former manager Mike Kane. His philosophy was "run it hard, fix it fast." In a way, he was the role model for those in the plant if they wished to remain employed. They may not have believed in his approach, but as Mike had said numerous times, "it is my way of the highway."

Using Figure 5-6, develop a strategy that will take you the new role model and the eastern plant manager from block 1 of Figure 5-6 through blocks #3, #2 and, ultimately, to block #4.

Your Migration Strategy

From block #1 to block #3

From block #3 to block # 2

From block #2 to block #4

Exercise 5-5 The Role Model Web

Chapter 17 explains the Web of Cultural Change; it indicates that for each of the eight elements of change (the radial spokes of the diagram), there are sixteen questions to be answered to create the score for that element. The second set of four questions are those that address role models. By answering these four specific questions for each of the eight elements of change, you can build a web diagram specific to role models.

This exercise is divided into three parts. The first part asks you to answer the role models questions for each of the eight elements of change. In Part 2, you use your scores from Part 1 to create your role models web of change. Then in Part 3, you analyze the web diagram by selecting those elements that received a low score and asking yourself (or your work group) why these scores were as low as they appeared. Chapter 18 in the text can help you accomplish this analysis.

Exercise 5-5 The Role Model Web, Part 1

Leadership **Total Score** ___________

5 Role models within the organization are recognized by those in leadership positions.

A. Strongly agree	5
B. Agree	4
C. Neutral	3
D. Disagree	2
E. Strongly disagree	1

6 The site role models support the change and new ways for conducting the work.

A. Strongly agree	5
B. Agree	4
C. Neutral	3
D. Disagree	2
E. Strongly disagree	1

7 Leadership clearly understands the impact that role models have on the work process, specifically a process that is undergoing change.

A. Strongly agree	5
B. Agree	4
C. Neutral	3
D. Disagree	2
E. Strongly disagree	1

8 Role models who do not support the change effort are reassigned or removed from their area of influence.

A. Strongly agree	5
B. Agree	4
C. Neutral	3
D. Disagree	2
E. Strongly disagree	1

Work Process **Total Score** ________

5 The role models have actively supported the work process design efforts that focus the organization on improved plant reliability.

A. Strongly agree	5
B. Agree	4
C. Neutral	3
D. Disagree	2
E. Strongly disagree	1

6 The role models have actively supported new and improved work processes when they are rolled out and implemented within the plant.

A. Strongly agree	5
B. Agree	4
C. Neutral	3
D. Disagree	2
E. Strongly disagree	1

7 The role model's behavior follows the work process as designed and implemented. They "walk the talk."

A. Strongly agree	5
B. Agree	4
C. Neutral	3
D. Disagree	2
E. Strongly disagree	1

8 Role models who are not in support of the work process are identified and corrective actions are taken. These actions could be coaching or other forms.

A. Strongly agree	5
B. Agree	4
C. Neutral	3
D. Disagree	2
E. Strongly disagree	1

Structure **Total Score** __________

5 The structural changes were made with involvement from the leadership team and role model. It was not a case of "here is how it is going to be done," another mandate from management.

A. Strongly agree	5
B. Agree	4
C. Neutral	3
D. Disagree	2
E. Strongly disagree	1

6 People in leadership positions within the structure are now or are becoming the role models for the organization.

A. Strongly agree	5
B. Agree	4
C. Neutral	3
D. Disagree	2
E. Strongly disagree	1

7 The current role models within the structure promote the new way things are to be done.

A. Strongly agree	5
B. Agree	4
C. Neutral	3
D. Disagree	2
E. Strongly disagree	1

8 Structural vacancies are filled with the objective of hiring individuals who will ultimately become role models and promote the new process.

A. Strongly agree	5
B. Agree	4
C. Neutral	3
D. Disagree	2
E. Strongly disagree	1

Group Learning **Total Score** ________

5 Role models who understand the organization's need to learn are in place.

A. Strongly agree	5
B. Agree	4
C. Neutral	3
D. Disagree	2
E. Strongly disagree	1

6 The role models actively support the learning process and, where it is not working, they encourage it.

A. Strongly agree	5
B. Agree	4
C. Neutral	3
D. Disagree	2
E. Strongly disagree	1

7 Role models recognize the need to use the Goal Achievement Model (or similar tool) to promote learning while, at the same time, accomplishing the vision, goals, and initiatives.

A. Strongly agree	5
B. Agree	4
C. Neutral	3
D. Disagree	2
E. Strongly disagree	1

8 When new things are learned that could have a positive effect on the business of reliability and maintenance, role models are open to trying them out instead of throwing up barriers that block new ideas.

A. Strongly agree	5
B. Agree	4
C. Neutral	3
D. Disagree	2
E. Strongly disagree	1

Technology **Total Score** ________

5 When role models require information to support the work, the technology makes it available.

A. Strongly agree	5
B. Agree	4
C. Neutral	3
D. Disagree	2
E. Strongly disagree	1

6 The technology provides reports related to the health of the work process; they are easily obtained and can be customized to the needs of the users.

A. Strongly agree	5
B. Agree	4
C. Neutral	3
D. Disagree	2
E. Strongly disagree	1

7 Role models actively support technology solutions in support of business needs.

A. Strongly agree	5
B. Agree	4
C. Neutral	3
D. Disagree	2
E. Strongly disagree	1

8 Role models promote data integration (a single pathway to all required information available to those who need it) instead of functional / technological information silos.

A. Strongly agree	5
B. Agree	4
C. Neutral	3
D. Disagree	2
E. Strongly disagree	1

Communication **Total Score** __________

5 Role models are good communicators of the organizational values.

A. Strongly agree	5
B. Agree	4
C. Neutral	3
D. Disagree	2
E. Strongly disagree	1

6 Role models are good listeners. They receive feedback that could either validate that the new process is understood or establish the need for clarification.

A. Strongly agree	5
B. Agree	4
C. Neutral	3
D. Disagree	2
E. Strongly disagree	1

7 Role models provide continuous communication, not just one time or with minimal frequency.

A. Strongly agree	5
B. Agree	4
C. Neutral	3
D. Disagree	2
E. Strongly disagree	1

8 Role models make themselves available to the organization to be good communicators. They view it as a key component of their work responsibilities.

A. Strongly agree	5
B. Agree	4
C. Neutral	3
D. Disagree	2
E. Strongly disagree	1

Interrelationships **Total Score** ________

5 Role models build organizational trust via positive interrelationships.

A. Strongly agree 5
B. Agree 4
C. Neutral 3
D. Disagree 2
E. Strongly disagree 1

6 Role models employ the concept of reciprocity in a positive manner.

A. Strongly agree 5
B. Agree 4
C. Neutral 3
D. Disagree 2
E. Strongly disagree 1

7 Role models create and support their allies in an effort to support the change process.

A. Strongly agree 5
B. Agree 4
C. Neutral 3
D. Disagree 2
E. Strongly disagree 1

8 Role models maintain both horizontal and vertical positive interrelationships.

A. Strongly agree 5
B. Agree 4
C. Neutral 3
D. Disagree 2
E. Strongly disagree 1

Rewards **Total Score** ________

5 The management team recognizes that money is not a long-term motivational reward for work well done. Alternative rewards are provided.

A. Strongly agree	5
B. Agree	4
C. Neutral	3
D. Disagree	2
E. Strongly disagree	1

6 The time delay between successful action and its reward is minimized wherever possible.

A. Strongly agree	5
B. Agree	4
C. Neutral	3
D. Disagree	2
E. Strongly disagree	1

7 Role models who promote the work process and other change efforts are visibly rewarded by the organization so that others can see what success looks like.

A. Strongly agree	5
B. Agrcc	4
C. Neutral	3
D. Disagree	2
E. Strongly disagree	1

8 Role models who do not promote the change (or may even work against it) visibly receive neutral or negative rewards so that the workforce can see what failure looks like.

A. Strongly agree	5
B. Agree	4
C. Neutral	3
D. Disagree	2
E. Strongly disagree	1

Exercise 5-5 The Role Model Web, Part 2

Enter the score from each of the elements into the table below.

Elements	Max Score	Baseline Score
Leadership	20	
Work Process	20	
Structure	20	
Group Learning	20	
Technology	20	
Communication	20	
Interrelationships	20	
Rewards	20	

Chart your scores on the web diagram to create your organizational value web.

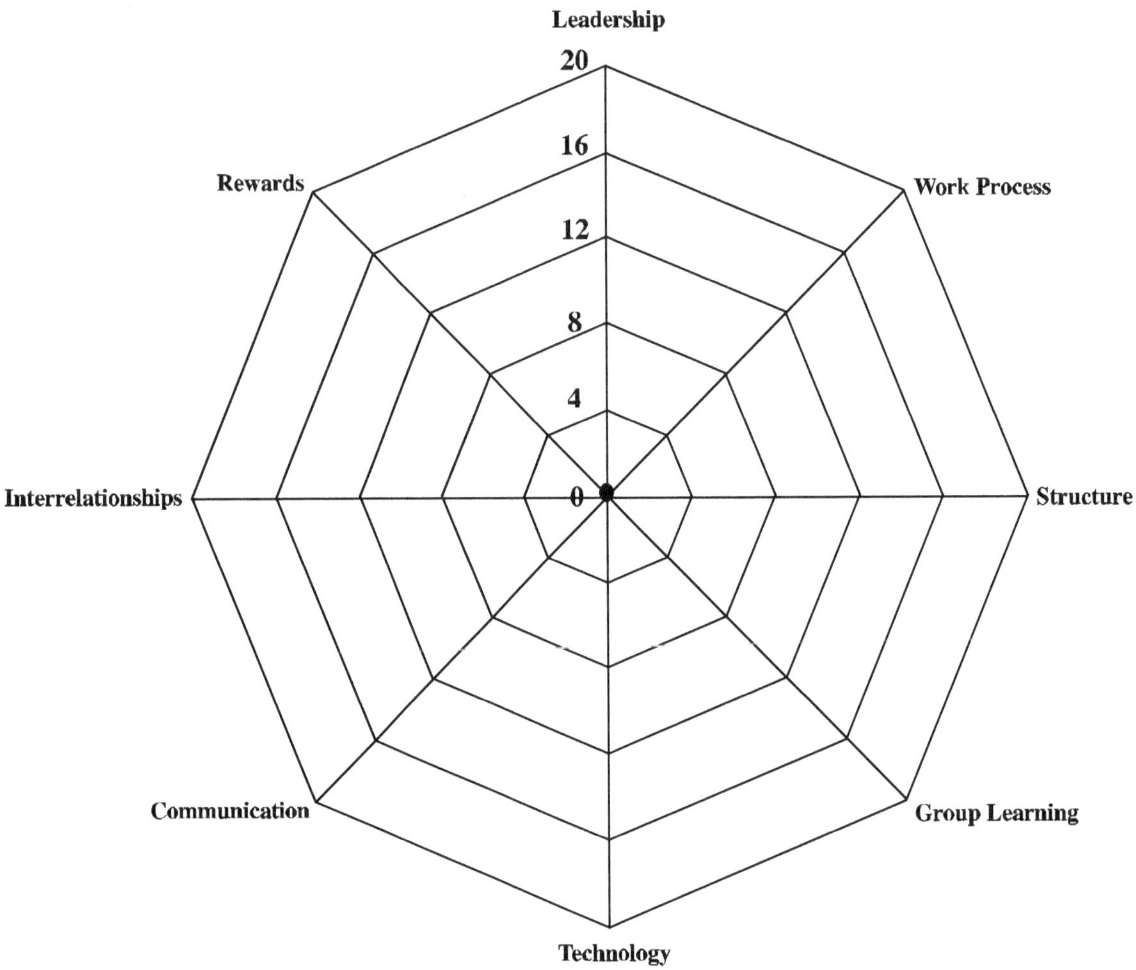

Exercise 5-5 The Role Model Web, Part 3

Explain why each of the eight elements of change related to role models has scored the way that it did. Describe what needs to be done to improve those scores that fall below 12.

Elements	Problem	Corrective Action
Leadership Score =		
Work Process Score =		
Structure Score =		
Group Learning Score =		
Technology Score =		
Communication Score =		
Interrelationships Score =		
Rewards Score =		

Figure 5-7 A Full-Size Role Model Web Diagram

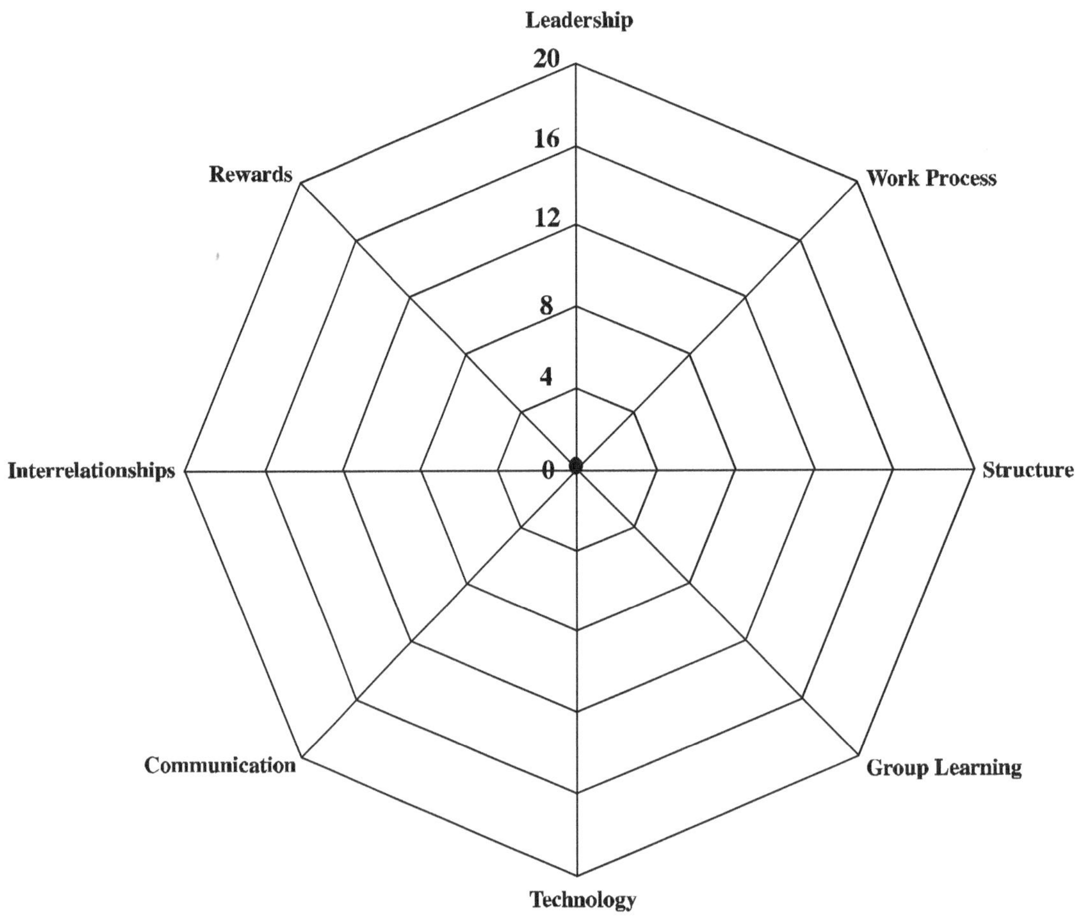

Rites and Rituals

Chapter 6

Purpose

This chapter explains concepts associated with the third of the four elements of culture: rituals and their supporting rites.

What Is Included

Figure 6-1 What Is a Ritual?

Figure 6-2 What Is a Rite?

Figure 6-3 Interaction Between Rites and Rituals

Figure 6-4 Rituals and Rites: The Reinforcing Loop

Figure 6-5 Changing Rituals and Their Supporting Rites

Exercise 6-1 Distinguishing Rites from Rituals

Exercise 6-2 Identifying Supporting Rites

Exercise 6-3 Correcting Rituals that Need Changing

Exercise 6-4 Rituals and Rites of Your Own

Exercise 6-5 Fixing Your Rituals and Their Supporting Rites

Exercise 6-6 The Rites and Rituals Web

Figure 6-6 A Full-Size Rites and Rituals Web Diagram

Figure 6-1 What Is a Ritual?

A ritual is a rule or set of rules
that guide our day-to-day work behavior.
Rituals are taken for granted because
they are an integral part of our jobs.
As they are repeated daily, rituals become an accepted
part of how business is conducted;
over time, they become invisible to those who follow them.
Yet they are extremely important
because they define what we do, how we do it,
and they represent the culture
and the value system in place in our plant.

Figure 6-2 What Is a Rite?

A rite is a company ceremony
or event that reinforces our rituals.
In a sense, rites provide a stage
for those involved to dramatize the culture
and organizational values to those in attendance.
Rituals and rites go hand in hand
because without accepted rituals, rites do not exist.

Figure 6-3 Interaction Between Rites and Rituals

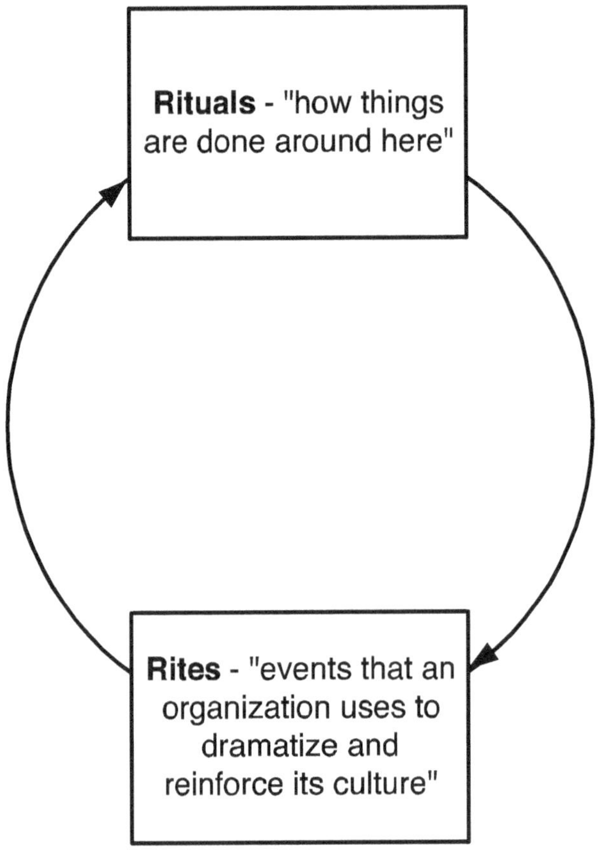

Figure 6-4 Rituals and Rites: The Reinforcing Loop

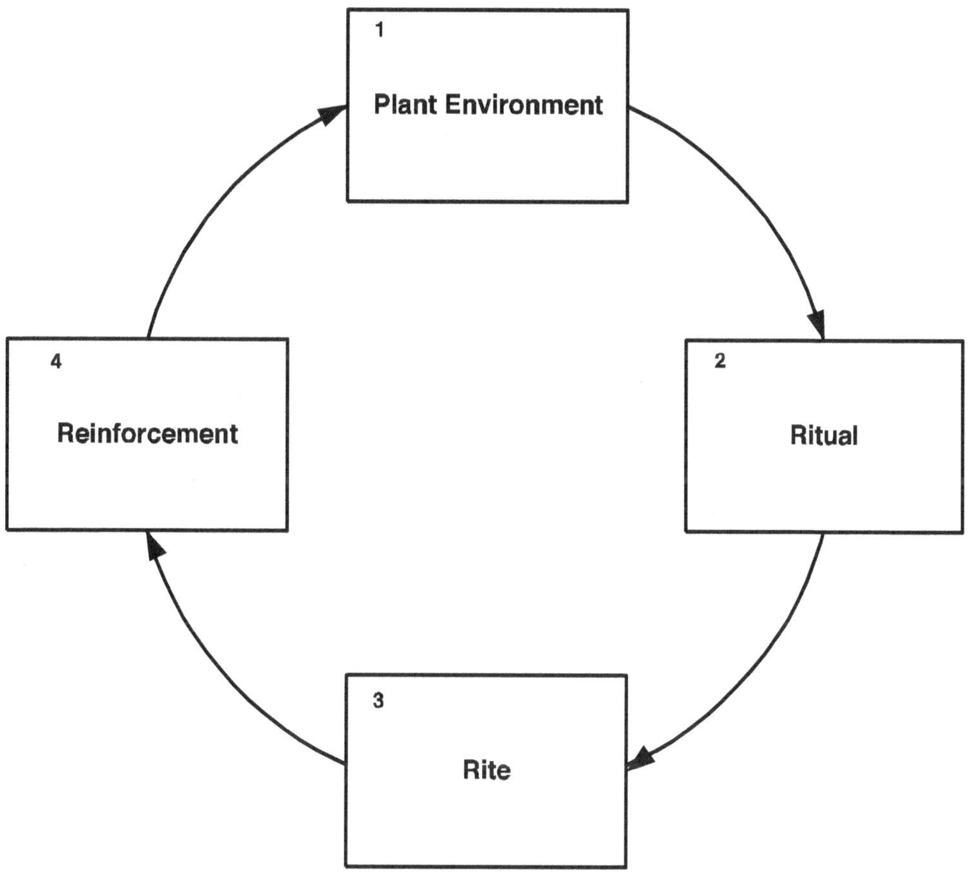

Figure 6-5 Changing Rituals and Their Supporting Rites

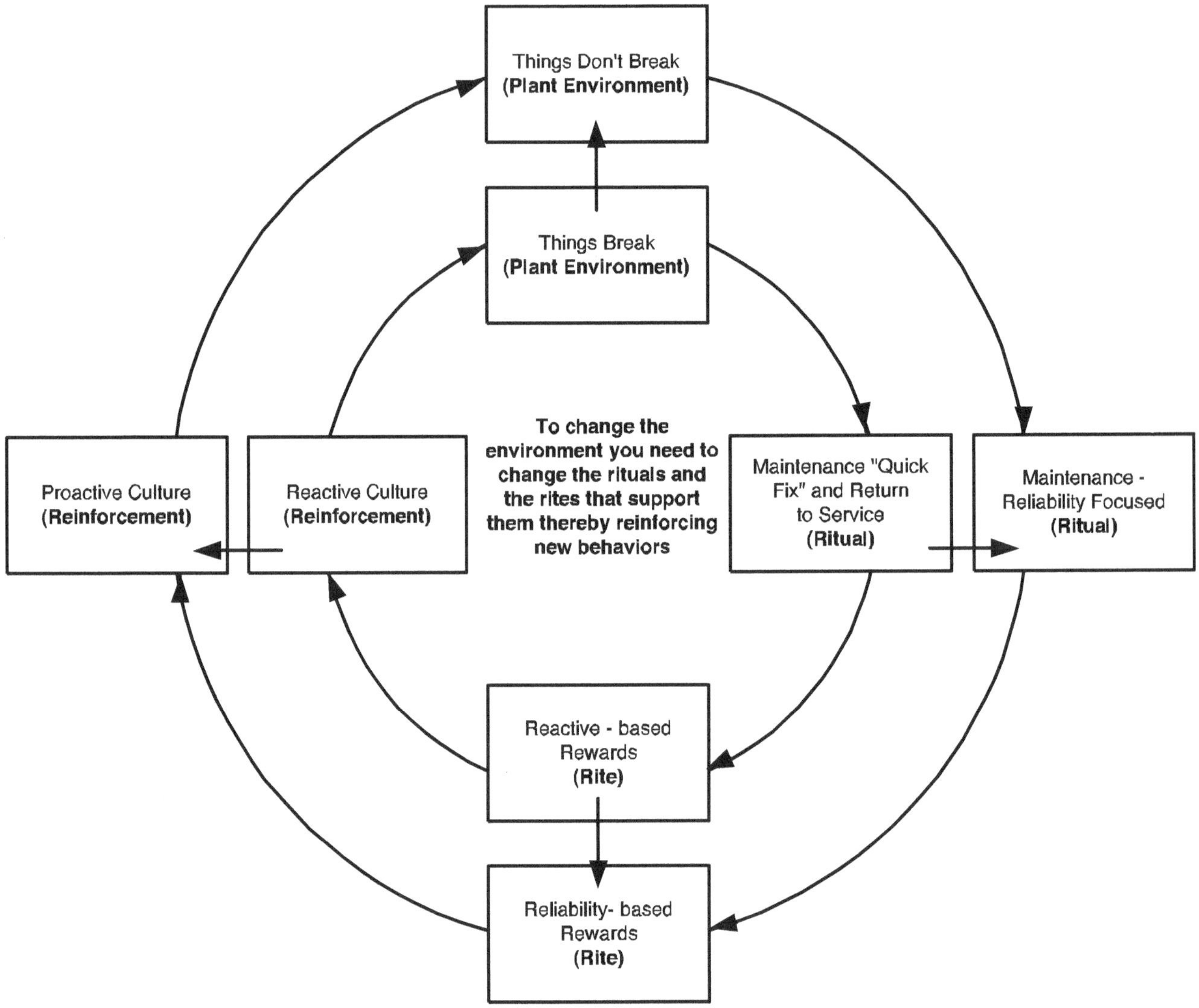

Exercise 6-1 Distinguishing Rites from Rituals

Different scenarios are provided in this exercise. Determine which ones you believe are rituals and which one you believe are rite. Check the appropriate column.

Scenario	Ritual	Rite
You and your work crew quickly fix a broken pump. The production manager says "nice job."		
All work in the plant in which you work is planned by full-time work planners.		
Each work crew is always assigned to the same production area.		
In recognition of twenty-five years of service, you get a luncheon with the plant manager.		
Safety tool box meetings are held with every crew every day.		
Yearly performance raises are awarded based on achieving the goals of the department.		
Your entire department just received a free lunch for achieving a safety goal.		
The day shift for the maintenance department 8:00 AM to 4:00 PM.		
For a job well done, you just received a one-time bonus check from your manager.		
Each time members of your crew do a poor job, they always seem to get assigned to the night shift.		

Exercise 6-2 Identifying Supporting Rites

In this exercise, various rituals are listed. What do you think the appropriate reinforcing rite would be for each of the listed rituals?

Ritual	Reinforcing Rite
The preventive maintenance program is alwayscompleted on schedule.	
The area mechanic comes in early just to make sure the line is functioning properly before the start of the shift.	
When production calls with a problem, maintenance is quick to respond.	
The maintenance department has done an excellent job of preventive and predictive maintenance; reliability has increased.	
When Joe got an emergency call on the weekend, he came into work to help solve the problem on his own time.	
The safety performance of the plant has increased from the prior year due to a great deal of work by the safety department.	
One of the engineers completes her Masters degree in engineering.	

Exercise 6-3 Correcting Rituals that Need Changing

If rites can reinforce rituals, they also can be used to alter them. Such rites don't reinforce the existing practices; they reinforce the change. Listed below are several rituals that are characteristic of a reactive organization. What rites would you employ to reinforce reliability-focused maintenance?

Reactive Ritual	What Rites Could Be Used to Alter it?
Whenever production calls on the phone, we immediately act on their request.	
When we make repairs, we seldom if ever call the reliability engineers to find out if there is something we could do to improve the equipment's performance.	
We invariably cancel our PM programmed work to do emergency work phoned in by production, even if it really isn't emergency in nature.	
When urgent work requests are received, the planners stop what they are doing and assist the foremen by coordinating material delivery.	
At the end of the day, the foremen hide the mobile equipment rather than returning them to the staging area for use by others. They do this in order to make sure they have them available if needed.	
Wearing personal protective safety equipment is a mandatory part of the plant's work rules. Nevertheless, it is often not enforced.	
The work crews always start late and quit early.	
Because most of the work in the plant is called an emergency, even though many are not, materials management always pays for expediting, without question.	

Exercise 6-4 Rituals and Rites of Your Own

Now that you understand the concept of rites and rituals, list some of your own.

Ritual	Reinforcing Rite	Is there a problem if you want to put a reliability - focused culture in place?

Exercise 6-5 Fixing Your Rituals and Their Supporting Rites

Select the rituals and the supporting rites that you would wish to change. Then work through the path to a solution by answering the questions below. Make a copy of this page so that you can identify and work on the solutions for more than one ritual and its supporting rites.

What is the ritual you selected?
Why do you want to change it? State the problem.
What are the reinforcing rites?
Why do the reinforcing rites undermine the change?
List the plant initiative that the current ritual supports.
If it does not support a plant initiative, does it undermine one? If so, which one?
If it does support a plant initiative, why is there a problem?
What needs to be done to change the ritual and its supporting rites?

Exercise 6-6 The Rites and Rituals Web

Chapter 17 explains the Web of Cultural Change; it indicates that for each of the eight elements of change (the radial spokes of the diagram), there are sixteen questions to be answered to create the score. The third set of four questions within each of the elements are those that address the rites and rituals. By answering these questions you can build a web diagram specific to the organizational rites and rituals.

This exercise is divided into three parts. The first part asks you to answer the rites and rituals questions for each of the eight elements of change. In Part 2, you use your scores from Part 1 to create your rites and rituals web of change. Finally in Part 3, you get the opportunity to analyze the web diagram by selecting those elements that received a low score and asking yourself (or your work group) why those scores were as low as they appeared. Chapter 18 in the text can help you accomplish this task.

Exercise 6-6 The Rites and Rituals Web, Part 1

Leadership **Total Score** ______

9 The site's rituals and their supporting rites are clearly understood.

A. Strongly agree	5
B. Agree	4
C. Neutral	3
D. Disagree	2
E. Strongly disagree	1

10 The rites and rituals that are out of sync with the new process are being altered.

A. Strongly agree	5
B. Agree	4
C. Neutral	3
D. Disagree	2
E. Strongly disagree	1

11 Old rites and rituals are not being reinforced.

A. Strongly agree	5
B. Agree	4
C. Neutral	3
D. Disagree	2
E. Strongly disagree	1

12 New rites and rituals are being reinforced by the site leadership and others.

A. Strongly agree	5
B. Agree	4
C. Neutral	3
D. Disagree	2
E. Strongly disagree	1

Work Process **Total Score** ________

9 The rituals that existed in the former work process have been altered or removed by those that support the new process.

A. Strongly agree	5
B. Agree	4
C. Neutral	3
D. Disagree	2
E. Strongly disagree	1

10 The supporting rites (for the rituals referenced in item #9) have been altered or removed to support the new process.

A. Strongly agree	5
B. Agree	4
C. Neutral	3
D. Disagree	2
E. Strongly disagree	1

11 Field review of the process has shown that the former rites and rituals have been removed from the process.

A. Strongly agree	5
B. Agree	4
C. Neutral	3
D. Disagree	2
E. Strongly disagree	1

12 When a rite or ritual that supports the old process is identified (you can't always identify them all at the outset), corrective action is taken to alter or remove them.

A. Strongly agree	5
B. Agree	4
C. Neutral	3
D. Disagree	2
E. Strongly disagree	1

Structure **Total Score** ________

9 The structure supports the rites and rituals as dictated by the work process.

A. Strongly agree 5

B. Agree 4

C. Neutral 3

D. Disagree 2

E. Strongly disagree 1

10 Roles and responsibilities are clear for everyone within the structure, specifically how they support the rites and rituals associated with the process.

A. Strongly agree 5

B. Agree 4

C. Neutral 3

D. Disagree 2

E. Strongly disagree 1

11 The former set of rites and rituals (those being replaced with the new process) have been removed from how things are done around here and audits are conducted to assure this fact.

A. Strongly agree 5

B. Agree 4

C. Neutral 3

D. Disagree 2

E. Strongly disagree 1

12 The work locations of those in the organizational structure support the rites and rituals of the process.

A. Strongly agree 5

B. Agree 4

C. Neutral 3

D. Disagree 2

E. Strongly disagree 1

Group Learning **Total Score** __________

9 Rites and rituals are in place and support the learning spiral.

A. Strongly agree 5

B. Agree 4

C. Neutral 3

D. Disagree 2

E. Strongly disagree 1

10 Rites and rituals are focused on supporting a learning organization, not a blaming one.

A. Strongly agree 5

B. Agree 4

C. Neutral 3

D. Disagree 2

E. Strongly disagree 1

11 Rites and rituals have embedded within them processes to identify gaps, learn from those gaps, and take the necessary corrective action.

A. Strongly agree 5

B. Agree 4

C. Neutral 3

D. Disagree 2

E. Strongly disagree 1

12 Rites are used as a learning and reinforcing process for the rituals. When necessary, rituals are modified based on what was learned.

A. Strongly agree 5

B. Agree 4

C. Neutral 3

D. Disagree 2

E. Strongly disagree 1

Technology **Total Score** ________

9 Technology supports the rituals of the site (process).

A. Strongly agree 5

B. Agree 4

C. Neutral 3

D. Disagree 2

E. Strongly disagree 1

10 Technology supports the rites that in turn support the rituals.

A. Strongly agree 5

B. Agree 4

C. Neutral 3

D. Disagree 2

E. Strongly disagree 1

11 Rituals exist within the process to permit acquisition and use of new software without having to go through large amounts of red tape.

A. Strongly agree 5

B. Agree 4

C. Neutral 3

D. Disagree 2

E. Strongly disagree 1

12 Rituals exist to permit continuous upgrading of software products in order to stay current with the vendor's releases.

A. Strongly agree 5

B. Agree 4

C. Neutral 3

D. Disagree 2

E. Strongly disagree 1

Communication **Total Score** __________

9 Rites are clearly communicated and there is a process to check understanding.

A. Strongly agree 5

B. Agree 4

C. Neutral 3

D. Disagree 2

E. Strongly disagree 1

10 Rituals are clearly communicated and there is a process to check understanding.

A. Strongly agree 5

B. Agree 4

C. Neutral 3

D. Disagree 2

E. Strongly disagree 1

11 The communications regarding rites and rituals have a lot of "why we are doing this and what are the benefits" content.

A. Strongly agree 5

B. Agree 4

C. Neutral 3

D. Disagree 2

E. Strongly disagree 1

12 Communication is widely used to reinforce the work process, the rituals, and their supporting rites.

A. Strongly agree 5

B. Agree 4

C. Neutral 3

D. Disagree 2

E. Strongly disagree 1

Interrelationships **Total Score** ________

9 Horizontal interrelationships are a very important aspect of the site rituals.

A. Strongly agree	5
B. Agree	4
C. Neutral	3
D. Disagree	2
E. Strongly disagree	1

10 Vertical interrelationships are a very important aspect of the site rituals.

A. Strongly agree	5
B. Agree	4
C. Neutral	3
D. Disagree	2
E. Strongly disagree	1

11 Rituals that exist were built so that positive interrelationships are a result.

A. Strongly agree	5
B. Agree	4
C. Neutral	3
D. Disagree	2
E. Strongly disagree	1

12 Rites that exist (in support of rituals) were built so that positive interrelationships are promoted as a result.

A. Strongly agree	5
B. Agree	4
C. Neutral	3
D. Disagree	2
E. Strongly disagree	1

Rewards **Total Score** __________

9 Where appropriate, rituals carry timely rewards that reinforce the change effort.

A. Strongly agree	5
B. Agree	4
C. Neutral	3
D. Disagree	2
E. Strongly disagree	1

10 Rites are designed to provide rewards that reinforce the rituals and the process.

A. Strongly agree	5
B. Agree	4
C. Neutral	3
D. Disagree	2
E. Strongly disagree	1

11 Where rewards do not reinforce the change effort, they are altered.

A. Strongly agree	5
B. Agree	4
C. Neutral	3
D. Disagree	2
E. Strongly disagree	1

12 Where there is a time delay that will diminish the value of a reward, an active effort is made to eliminate the delay or make it as small as possible.

A. Strongly agree	5
B. Agree	4
C. Neutral	3
D. Disagree	2
E. Strongly disagree	1

Exercise 6-6 The Rites and Rituals Web, Part 2

Enter the score from each of the elements into the table below.

Elements	Max Score	Baseline Score
Leadership	20	
Work Process	20	
Structure	20	
Group Learning	20	
Technology	20	
Communication	20	
Interrelationships	20	
Rewards	20	

Chart your scores on the web diagram to create your organizational value web.

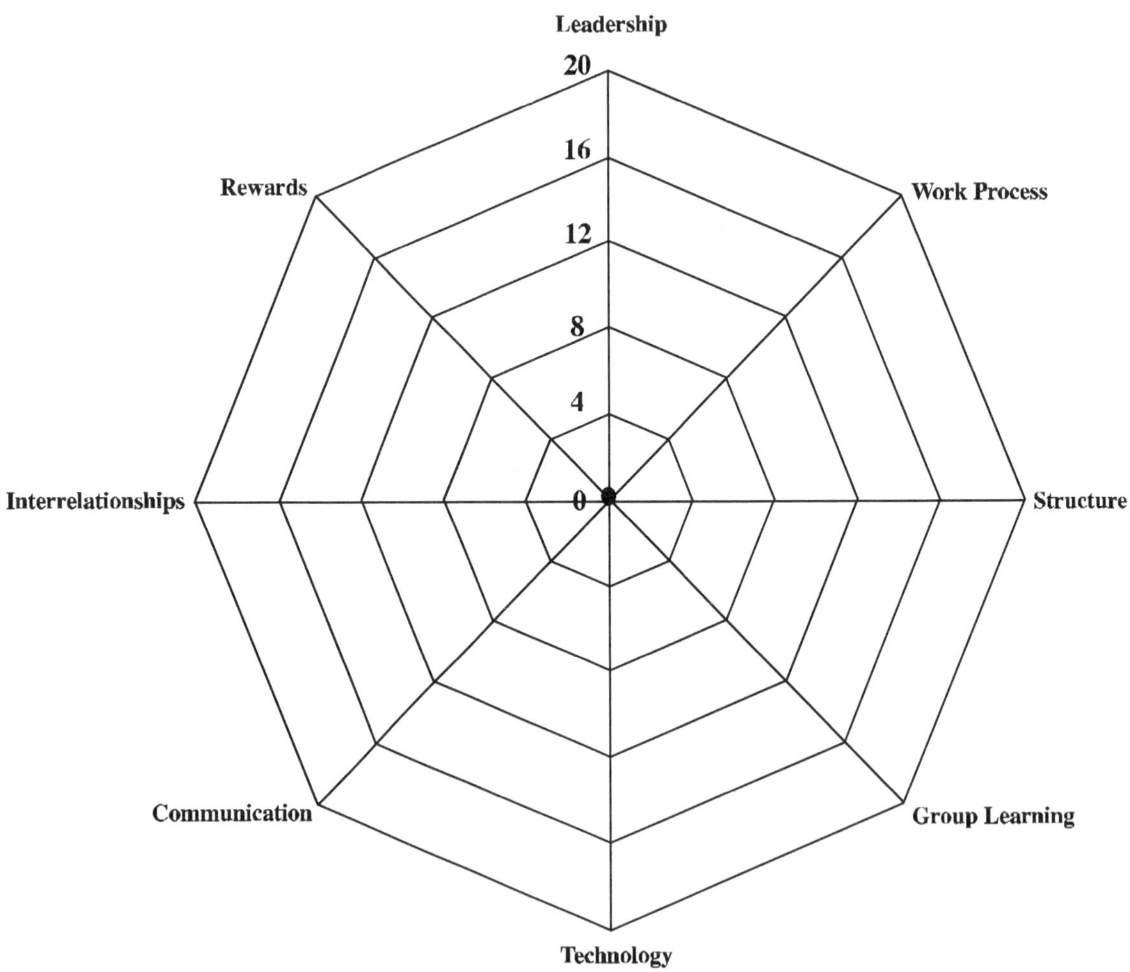

Exercise 6-6 The Rites and Rituals Web, Part 3

Explain why each of the eight elements of change related to organizational values has scored the way that it did. Describe what needs to be done to improve those scores that fall below 12.

Elements	Problem	Corrective Action
Leadership Score =		
Work Process Score =		
Structure Score =		
Group Learning Score =		
Technology Score =		
Communication Score =		
Interrelationships Score =		
Rewards Score =		

Figure 6-6 A Full-Size Rites and Rituals Web Diagram

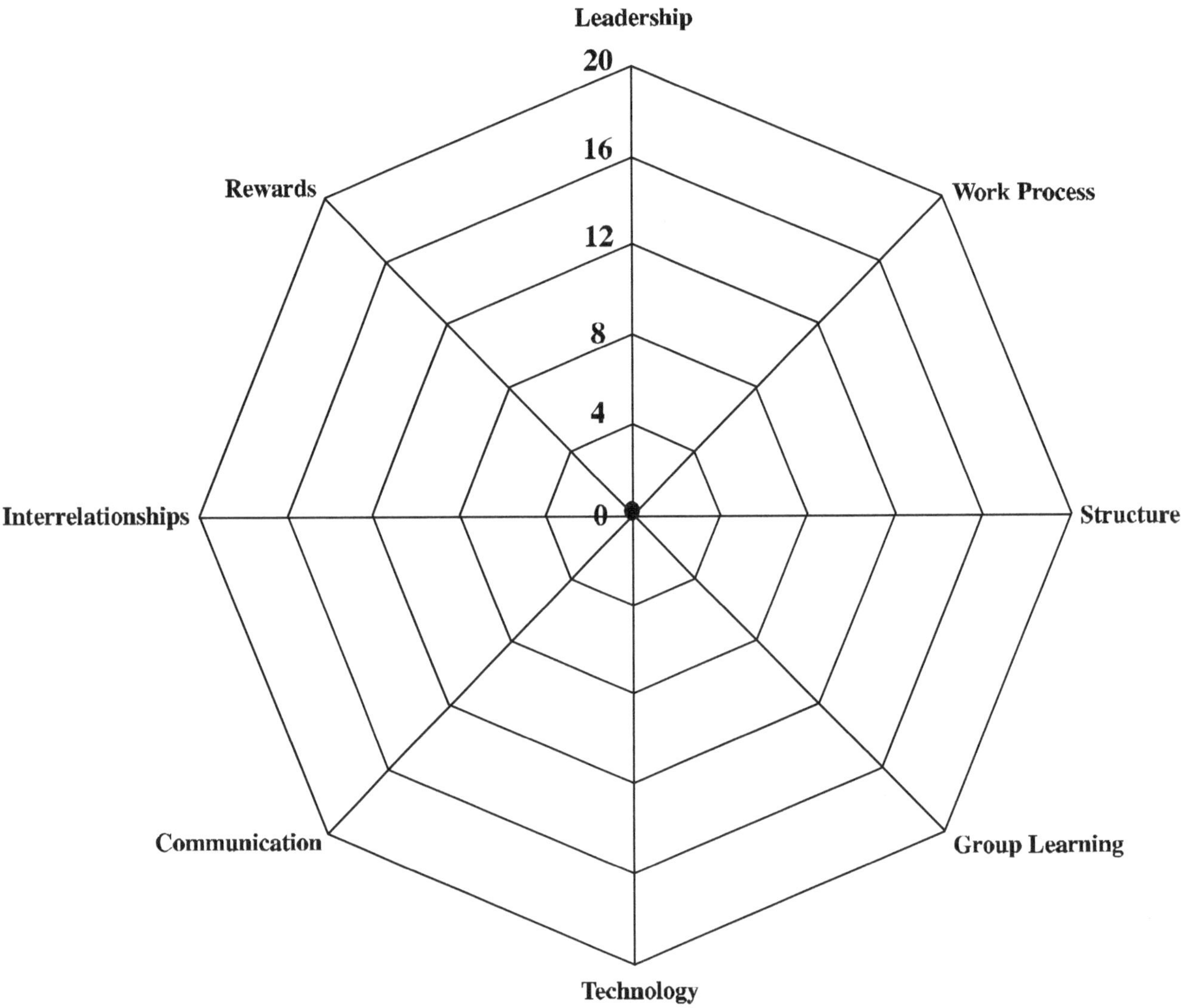

Cultural Infrastructure

Chapter 7

Purpose

This chapter introduces the last of the four elements of cultural change: cultural infrastructure. An understanding of this people-oriented component is critical for successful change to take place.

What Is Included

Figure 7-1	Cultural Infrastructure Defined
Figure 7-2	Cultural Infrastructure Components Defined
Exercise 7-1	Identifying the Cultural Infrastructure
Figure 7-3	How to Work with Cultural Infrastructure Components
Exercise 7-2	Story Tellers
Exercise 7-3	Keepers of the Faith
Exercise 7-4	Whisperers
Exercise 7-5	Gossips
Exercise 7-6	Spies
Exercise 7-7	Symbols
Exercise 7-8	Language
Exercise 7-9	The Cultural Infrastructure Web
Figure 7-4	A Full-Size Cultural Infrastructure Web Diagram

Figure 7-1 Cultural Infrastructure Defined

Cultural infrastructure is the hidden hierarchy
of people and communication processes
that binds the organization
to the culture
and sustains it over time.

Figure 7-2 Cultural Infrastructure Components Defined

Story Tellers –	promoting the culture through war stories
Keepers of the Faith –	mentors and protectors of the culture
Whisperers –	passers of information behind the scenes
Gossips –	the hidden day-to-day communication system
Spies –	passers of sensitive information to those who may or may not need to know
Symbols –	mechanisms for conveying what and who is important
Language –	terminology that describes what is done and often how

Exercise 7-1 Identifying the Cultural Infrastructure

As we have learned the cultural infrastructure has various components. For each scenario listed below, identify which of the cultural infrastructure components is being described.

	Scenario	Infrastructure Type
1	You are at a meeting proposing a new reliability work strategy. Before you get started, Pat begins to relate a story about how the old process saved the plant several years ago.	
2	You have developed a new approach to pump preventive maintenance. One of the senior engineers who has been with the firm a very long time takes you aside and explains why the existing process is better.	
3	At your daily maintenance meeting, you and your work team are very open and honest with each other about the work process. Successes and failures are openly discussed and analyzed to see where improvements could be made. For some strange reason, your boss always knows of your failures.	
4	During a critical repair effort, you recognize that you neglected to order a very important part. This is quite embarrassing, but by working with the vendor you have it delivered on time. Within a day, everyone in your department knows of your mistake.	
5	You casually mention to a co-worker a new idea that you believe would help the work team; however, it still needs development. Later that day, your boss stops by to talk. From the discussion, you recognize she knows about the idea.	

	Scenario	Infrastructure Type
6	You and two other teams are working on the same reliability problem. Your supervisor has set up this competition to have the organization develop multiple and independent approaches to the problem. When presentations are made, it is obvious that one of the other teams has used your concept.	
7	You are moving to a new office building. In your former building, people occupied whatever office they were assigned without regard to their job grade. In the new building, each job grade gets an office of a specified size and location.	
8	You attend a meeting representing your department. Everyone else there is from the computer group. During the discussion, they use so many acronyms that you have no idea of what is being said.	
9	You recognize that there is a very serious problem in your organization, but you can't go to your supervisor because they are the root cause. You mention the problem to the manager's administrative assistant as part of your normal discussion. A few weeks later your supervisor is fired.	
10	As a foreman, you identify several inefficient work practices. As a result you develop some new ideas. However, when you share them with your supervisor you are told not to rock the boat.	

Figure 7-3 How to Work with Cultural Infrastructure Components

The Cultural Infrastructure

Components	What to Do	How to Do It
Storytellers	Engage them in the change process.	Get the storytellers to tell stories of the new way vs. that of the old.
Keepers of the Faith	Engage them in the change process.	Have them provide counseling and mentoring, but focused on the new process.
Whisperers	Engage them in the change process.	Work with them so that they filter and pass information supporting the change.
Gossips	Negate the value of their communication.	Provide upfront and continuous communication to reduce the material to gossip about.
Spies	Negate the value of their communication.	Same as the gossips, but it will undermine the value of the information they have to pass.
Symbols	Change to reflect the new reliability focus.	Alter the symbols to reflect the new focus. Reward those who contribute to the new process.
Language	Change to reflect the new reliability focus.	Provide education and involvement so that everyone speaks the same language.

Exercise 7-2 Story Tellers

Story tellers and their stories relate information about organizational values, the rituals and rites that were employed, and the role models who accomplished them. The stories try to perpetuate the organization's culture.

Who in your organization are the story tellers (no names)?
What is the evidence that substantiates this classification?
How can what they do inhibit change?
How could you manage the negative aspects?
How could you utilize the story tellers (and the stories they tell) to promote change?

Exercise 7-3 Keepers of the Faith

Keepers of the Faith are the coaches and mentors of the organization. We go to them for guidance when we have a question about how to proceed. They may be the firm's managers or they simply may be individuals who have been with the company for a long time and know how things are done. What is important is that when we solicit their council, they provide the basics of how things get done within the context of the organization's culture.

Who in your organization are the Keepers of the Faith (no names)?
What is the evidence that substantiates this classification?
How can what they do inhibit change?
How could you manage the negative aspects?
How could you utilize them to promote change?

Exercise 7-4 Whisperers

Whisperers are those members of the organization who have the ability to whisper in the boss's ear, hence their name. Whispering is their source of power because they are not typically leaders of the organization. However, they play an important role within the cultural infrastructure by conveying information upwards that may not otherwise be communicated within normal channels

Who in your organization are the whisperers (no names)?
What is the evidence that substantiates this classification?
How can what they do inhibit change?
How could you manage the negative aspects?
How could you utilize them to promote change?

Exercise 7-5 Gossips

Gossips are like whisperers, except they communicate to a different audience. Whereas the whisperers communicate upwards, the gossips communicate throughout the entire organization. They spread stories, rumors, and other sorts of information. They are part of the plant's informal communication system.

Who in your organization are the gossips (no names)?
What is the evidence that substantiates this classification?
How can what they do inhibit change?
How could you manage the negative aspects?
How could you utilize them to promote change?

Exercise 7-6 Spies

A spy is a person who obtains sensitive information by various and often devious methods, then passes it to people or groups who use it to gain advantage over their rivals or adversaries.

Who in your organization are the spies (no names)?
What is the evidence that substantiates this classification?
How does what they do inhibit change?
How could you manage or control this aspect of the cultural infrastructure?
How could you utilize them to promote change?

Exercise 7-7 Symbols

Symbols are that part of the culture that tells us who we are as well as how important the organization thinks we are in the scheme of things.

What symbols are used in your organization, department, or work team? Cite examples.
How do they inhibit change?
How could you manage the negative aspects?
How could you utilize symbols to promote the change process?

Exercise 7-8 Language

Part of the cultural infrastructure is composed of the language we use in our day-to-day communication with one another. This language is usually loaded with acronyms – initials that mean something to those who are part of the process being discussed. Understanding these acronyms clearly identifies who is part of the specific sub-culture. Those who are not knowledgeable in the sub-culture's language often do not understanding what is being said.

Does your organization, department, or work group use a specific language (acronyms, names, etc.) to communicate? Cite some examples.
How does the use of this language inhibit change?
How could you control it?
How could you utilize language to promote the change process?

Exercise 7-9 The Cultural Infrastructure Web

Chapter 17 explains the Web of Cultural Change; it indicates that for each of the eight elements of change (the radial spokes of the diagram), there are sixteen questions to be answered to create the score for that element. The last four questions address the cultural infrastructure. By answering these four specific questions, you can build a web diagram specific to the cultural infrastructure.

The first part of this exercise asks you to answer the cultural infrastructure questions for each of the eight elements of change. In Part 2, you use your scores from Part 1 to create your cultural infrastructure web of change. In Part 3, you analyze the web diagram by selecting the elements that have low scores and asking yourself why they were low. Chapter 18 can help you accomplish this task.

Exercise 7-9 The Cultural Infrastructure Web, Part 1

Leadership **Total Score** __________

13 Leadership understands the importance of addressing the cultural infrastructure as part of any change effort.

A. Strongly agree	5
B. Agree	4
C. Neutral	3
D. Disagree	2
E. Strongly disagree	1

14 The Keepers of the Faith have been involved with the development of the change initiative.

A. Strongly agree	5
B. Agree	4
C. Neutral	3
D. Disagree	2
E. Strongly disagree	1

15 Gossips, whisperers, and spies are identified and have been part of the process to some extent.

A. Strongly agree	5
B. Agree	4
C. Neutral	3
D. Disagree	2
E. Strongly disagree	1

16 Language and symbols have been developed to support the change effort.

A. Strongly agree	5
B. Agree	4
C. Neutral	3
D. Disagree	2
E. Strongly disagree	1

Work Process **Total Score** ________

13 The cultural infrastructure's language and symbols have evolved to incorporate the new process.

A. Strongly agree 5

B. Agree 4

C. Neutral 3

D. Disagree 2

E. Strongly disagree 1

14 The key members of the cultural infrastructure have been identified and their support elicited in developing and rolling out the new process.

A. Strongly agree 5

B. Agree 4

C. Neutral 3

D. Disagree 2

E. Strongly disagree 1

15 A sincere effort has been made to focus stories towards success of the new process vs. how well things were done under the old. The Story Tellers have been used to support this effort.

A. Strongly agree 5

B. Agree 4

C. Neutral 3

D. Disagree 2

E. Strongly disagree 1

16 An effective process is in place to assure that the cultural infrastructure's communication channels (used by the whisperers, gossips, and spies) are focused in support of the new process.

A. Strongly agree 5

B. Agree 4

C. Neutral 3

D. Disagree 2

E. Strongly disagree 1

Structure **Total Score** __________

13 The members of the cultural infrastructure whose roles have been altered by the change have been identified, and efforts have been made to smooth this transition.

A. Strongly agree 5

B. Agree 4

C. Neutral 3

D. Disagree 2

E. Strongly disagree 1

14 A structural change plan was coupled with effective and efficient communication and used to minimize the effect of the gossips.

A. Strongly agree 5

B. Agree 4

C. Neutral 3

D. Disagree 2

E. Strongly disagree 1

15 A structural change plan was coupled with effective and efficient communication and used to minimize the effect of the whisperers.

A. Strongly agree 5

B. Agree 4

C. Neutral 3

D. Disagree 2

E. Strongly disagree 1

16 The Keepers of the Faith have been included in the structural change process and enlisted to support it by their cultural infrastructure involvement.

A. Strongly agree 5

B. Agree 4

C. Neutral 3

D. Disagree 2

E. Strongly disagree 1

Group Learning **Total Score** ______

13 The cultural infrastructure is monitored to remove learning barriers.

A. Strongly agree	5
B. Agree	4
C. Neutral	3
D. Disagree	2
E. Strongly disagree	1

14 Active efforts are in place to promote group learning within the organization, especially within the cultural infrastructure.

A. Strongly agree	5
B. Agree	4
C. Neutral	3
D. Disagree	2
E. Strongly disagree	1

15 Language and symbols are being learned that represent the new process.

A. Strongly agree	5
B. Agree	4
C. Neutral	3
D. Disagree	2
E. Strongly disagree	1

16 The Keepers of the Faith are being enlisted to mentor the learning process within the site.

A. Strongly agree	5
B. Agree	4
C. Neutral	3
D. Disagree	2
E. Strongly disagree	1

Technology **Total Score** ________

13 Storytellers promote technology through their stories of successful implementations and the benefits delivered.

A. Strongly agree	5
B. Agree	4
C. Neutral	3
D. Disagree	2
E. Strongly disagree	1

14 Keepers of the Faith (mentors) promote the use of technology for sound reliability-based business decisions.

A. Strongly agree	5
B. Agree	4
C. Neutral	3
D. Disagree	2
E. Strongly disagree	1

15 Software and hardware issues are widely communicated to diminish the potential negatives that could be conveyed by the whisperers, gossips, and spies.

A. Strongly agree	5
B. Agree	4
C. Neutral	3
D. Disagree	2
E. Strongly disagree	1

16 There is a good communication plan to promote new technology through language and symbols.

A. Strongly agree	5
B. Agree	4
C. Neutral	3
D. Disagree	2
E. Strongly disagree	1

Communication **Total Score** __________

13	Storytellers' communications support the process instead of how it was done around here before the change.	
	A. Strongly agree	5
	B. Agree	4
	C. Neutral	3
	D. Disagree	2
	E. Strongly disagree	1
14	Keepers of the Faith (mentors) focus their communication more on the new process instead of trying to show how the old way was better.	
	A. Strongly agree	5
	B. Agree	4
	C. Neutral	3
	D. Disagree	2
	E. Strongly disagree	1
15	A large amount of communication keeps everyone informed, diminishing the value of the gossips.	
	A. Strongly agree	5
	B. Agree	4
	C. Neutral	3
	D. Disagree	2
	E. Strongly disagree	1
16	There is a large amount of communication keeps everyone informed, diminishing the value of the whisperers.	
	A. Strongly agree	5
	B. Agree	4
	C. Neutral	3
	D. Disagree	2
	E. Strongly disagree	1

Interrelationships **Total Score** __________

13 Storytellers promote positive interrelationships across the system by the stories that they relate.

A. Strongly agree	5
B. Agree	4
C. Neutral	3
D. Disagree	2
E. Strongly disagree	1

14 Keepers of the Faith (mentors) promote positive interrelationships as one of the best ways to get things done around here.

A. Strongly agree	5
B. Agree	4
C. Neutral	3
D. Disagree	2
E. Strongly disagree	1

15 Role models and other members of the management team promote positive interrelationships with the whisperers to enable themselves to get the real story about what is going on with the process.

A. Strongly agree	5
B. Agree	4
C. Neutral	3
D. Disagree	2
E. Strongly disagree	1

16 Role models and other members of the management team promote positive interrelationships with the gossips. This helps them find out what is going on with the process and, where necessary, correct miscommunications.

A. Strongly agree	5
B. Agree	4
C. Neutral	3
D. Disagree	2
E. Strongly disagree	1

Rewards **TotalScore** __________

13 Members of the cultural infrastructure recognize that one reward if they remain in roles that are key to the infrastructure is their support for the change effort.

A. Strongly agree	5
B. Agree	4
C. Neutral	3
D. Disagree	2
E. Strongly disagree	1

14 Management actively engages and rewards cultural infrastructure members for their help in promoting the change.

A. Strongly agree	5
B. Agree	4
C. Neutral	3
D. Disagree	2
E. Strongly disagree	1

15 Negative and / or neutral rewards are provided to those members of the cultural infrastructure who use their roles to disrupt the change process.

A. Strongly agree	5
B. Agree	4
C. Neutral	3
D. Disagree	2
E. Strongly disagree	1

16 Spies are not rewarded for passing sensitive information. At the minimum, they are informed that they are disrupting the process.

A. Strongly agree	5
B. Agree	4
C. Neutral	3
D. Disagree	2
E. Strongly disagree	1

Exercise 7-9 The Cultural Infrastructure Web, Part 2

Enter the score from each of the elements into the table below.

Elements	Max Score	Baseline Score
Leadership	20	
Work Process	20	
Structure	20	
Group Learning	20	
Technology	20	
Communication	20	
Interrelationships	20	
Rewards	20	

Chart your scores on the web diagram to create your cultural infrastructure web.

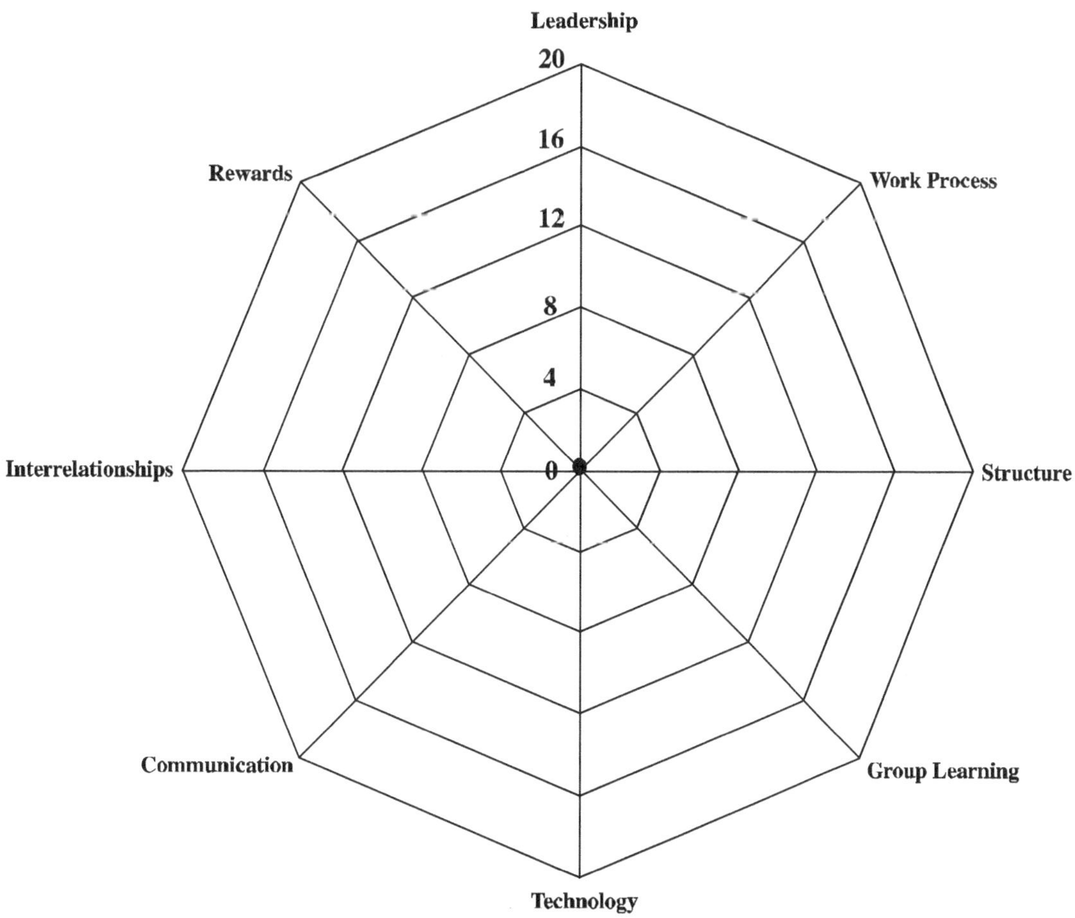

Exercise 7-9 The Cultural Infrastructure Web, Part 3

Explain why each of the eight elements of change related to cultural infrastructure has scored the way that it did. Describe what needs to be done to improve those scores that fall below 12.

Elements	Problem	Corrective Action
Leadership Score =		
Work Process Score =		
Structure Score =		
Group Learning Score =		
Technology Score =		
Communication Score =		
Interrelationships Score =		
Rewards Score =		

Figure 7-4 A Full-Size Cultural Infrastructure Web Diagram

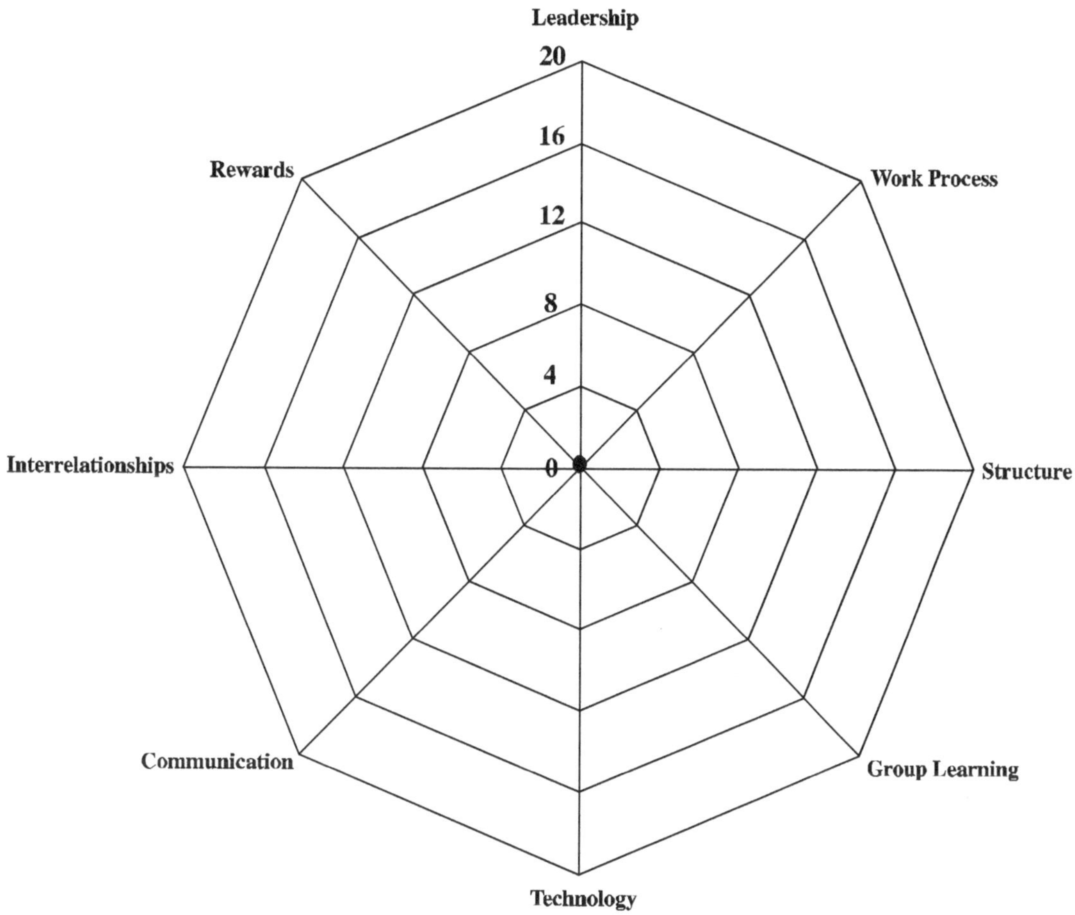

Chapter 8

The Elements of Change

Purpose

This chapter brings into sharp focus the interrelationships between the four elements of culture and the eight elements of change. For change to be successful, understanding this relationship is very important; missing elements cause holes in the change process that will undermine the overall effort.

What Is Included

Figure 8-1	The Relationship Between the Four Elements of Culture and the Eight Elements of Change
Figure 8-2	How the Eight Elements of Change Fit Together
Exercise 8-1	How Do the Eight Elements of Change Promote Change?
Figure 8-3	The Spokes of the Web of Change
Figure 8-4	What If One of the Eight Elements of Change Was Missing?
Exercise 8-2	The Impact of Missing One of the Elements of Change
Figure 8-5	The Web of Change with Culture at the Center
Figure 8-6	What If One of the Four Elements of Culture Was Missing?
Exercise 8-3	The Impact of Missing The Elements of Culture
Figure 8-7	The Relationship Between the Elements of Change and Culture
Figure 8-8	How Do the Elements of Change and Culture Impact Each Other?
Exercise 8-4	The Impact of the Elements on Each Other

Figure 8-1 The Relationship Between the Four Elements of Culture and the Eight Elements of Change

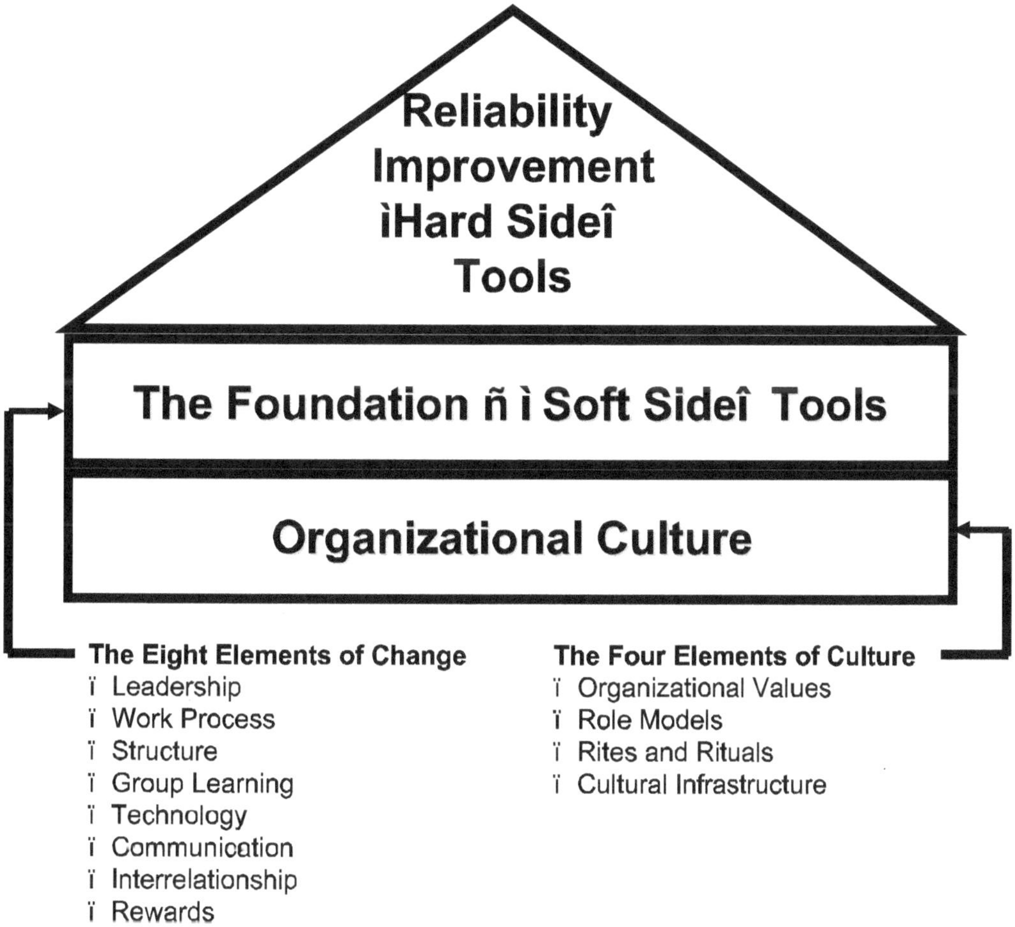

Figure 8-2 How the Eight Elements of Change Fit Together

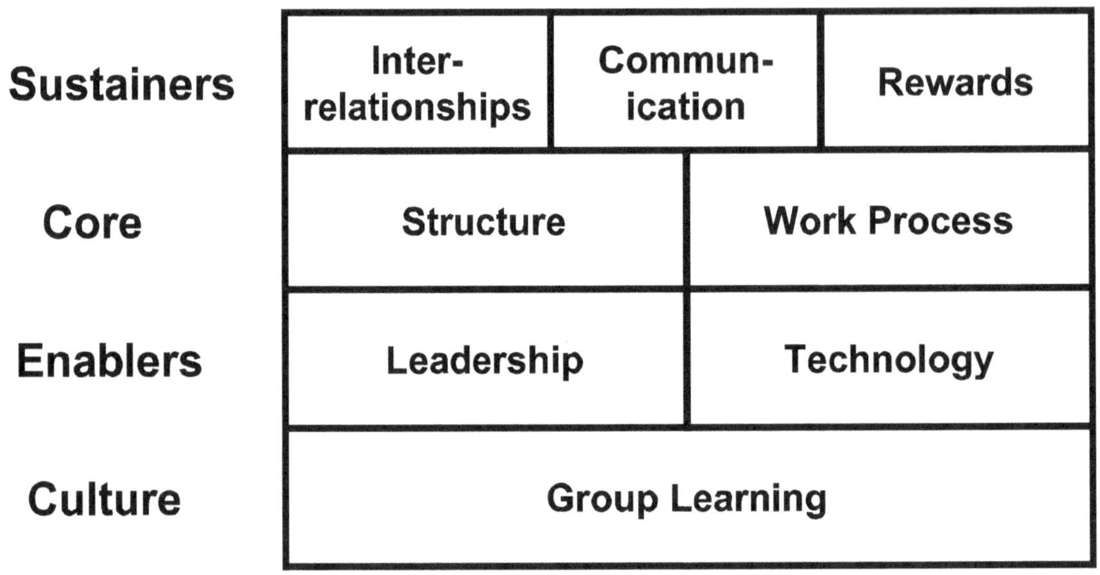

Exercise 8-1 How Do the Eight Elements of Change Promote Change?

Complete the following chart about successful change initiatives, relative to each of the eight elements of change. How does each of the elements and the role that each plays (as shown in Figure 8-2) contribute to the process of change? If possible, use a successful change initiative in which you were involved. If an example is not available, consider the elements in a general context.

Elements	How Does This Element Contribute to the Process of Change?
Leadership	
Work Process	
Structure	
Group Learning	
Technology	
Communication	
Interrelationships	
Rewards	

Figure 8-3 The Spokes of the Web of Change

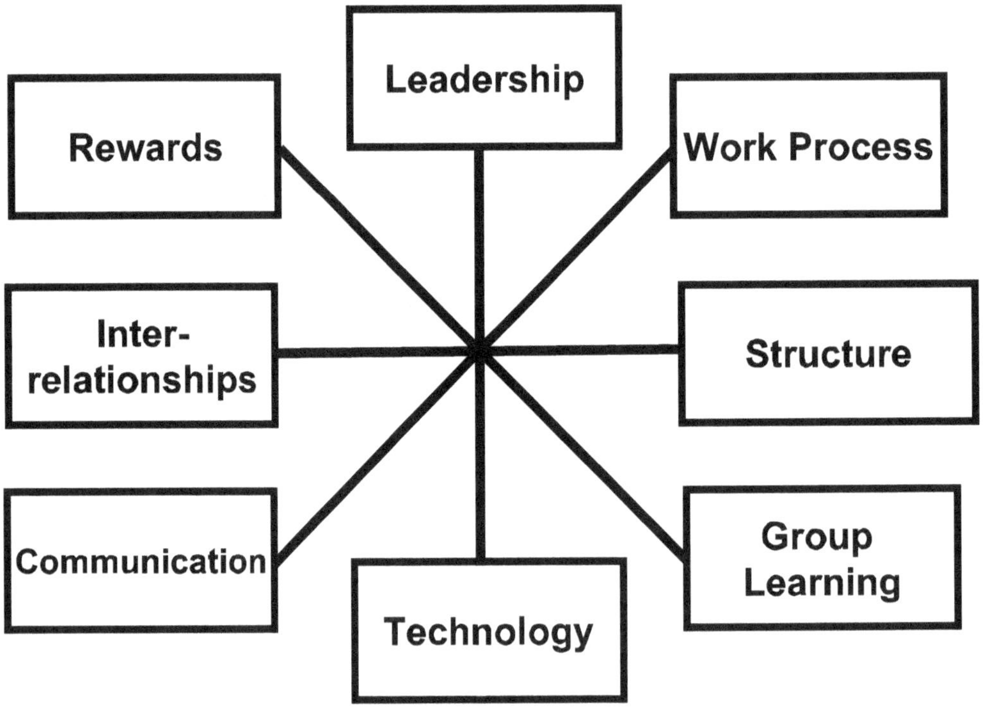

Figure 8-4 What If One of the Eight Elements of Change Was Missing?

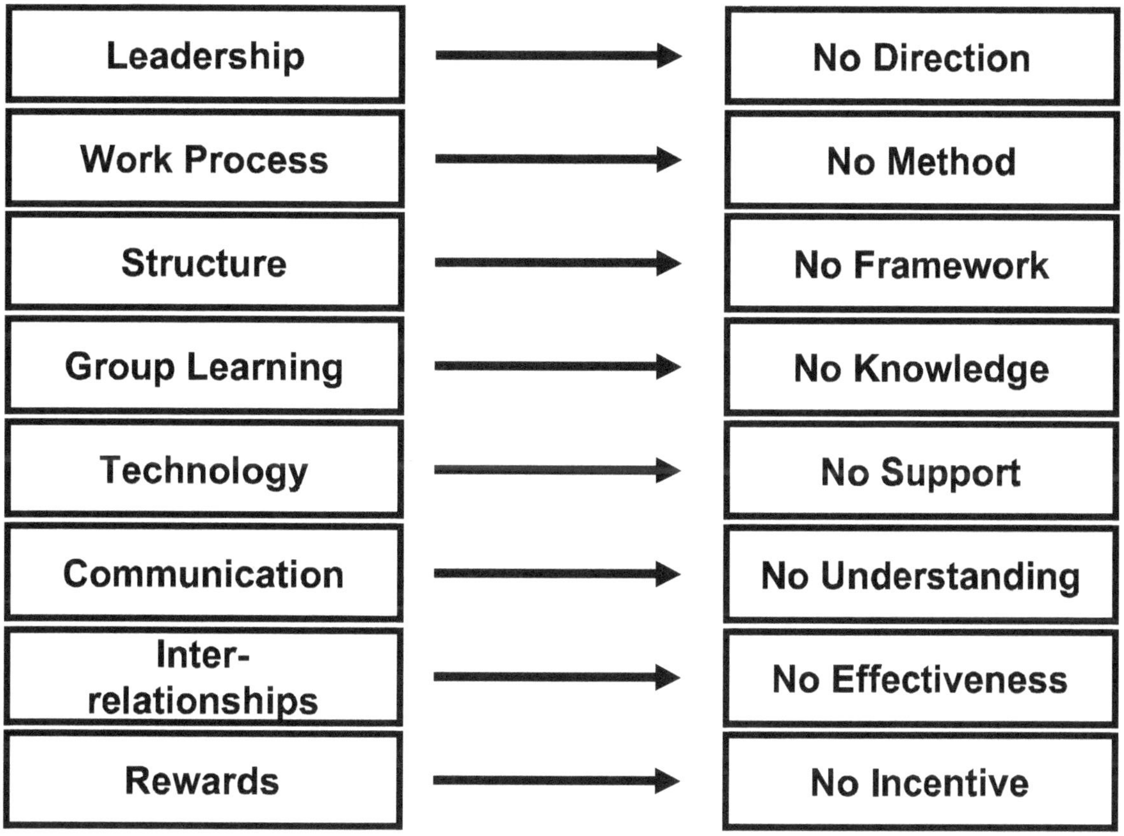

Exercise 8-2 The Impact of Missing One of the Elements of Change, Part 1

Select a successful change initiative in which you were involved. Then identify what you believe would have happened if one of the eight elements of change was missing. Complete this exercise for each of the elements.

Elements	What Would Have Been the Outcome If This Element Was Missed?
Leadership	
Work Process	
Structure	
Group Learning	
Technology	
Communication	
Interrelationships	
Rewards	

Exercise 8-2 The Impact of Missing One of the Elements of Change, Part 2

To get a different perspective, select an unsuccessful change initiative. Then identify which elements of change were missed. Explain what you believe the outcome would have been if this element was not missed and why.

Elements	X = missed	What Would Have Been the Outcome If This Element Was Not Missed?
Leadership		
Work Process		
Structure		
Group Learning		
Technology		
Communication		
Interrelationships		
Rewards		

Figure 8-5 The Web of Change with Culture at the Center

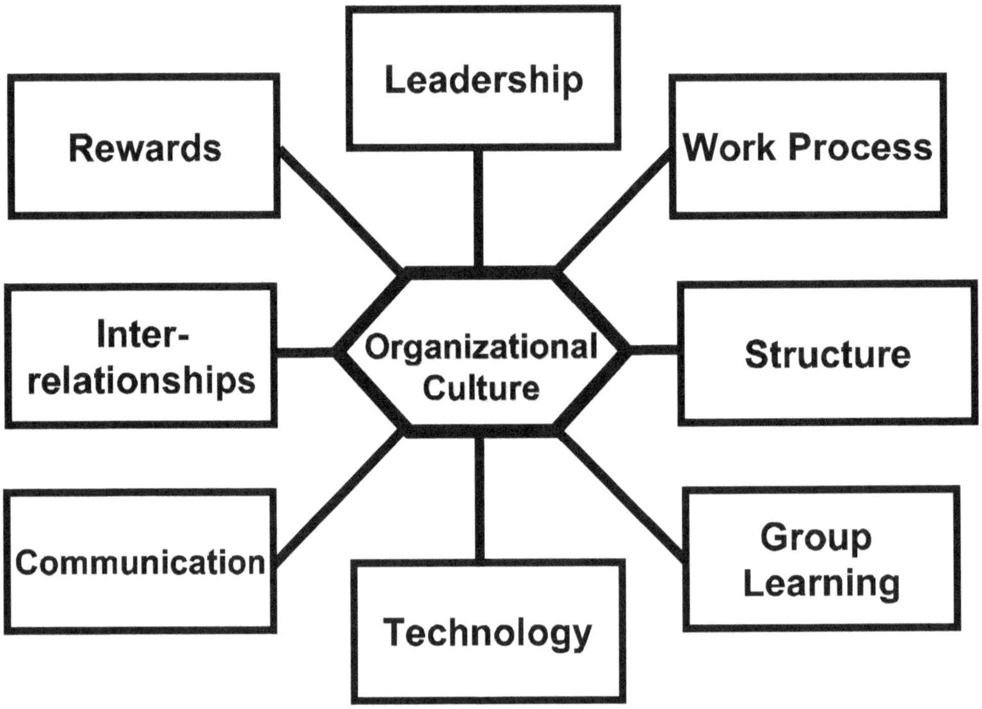

Figure 8-6 What If One of the Four Elements of Culture Was Missing?

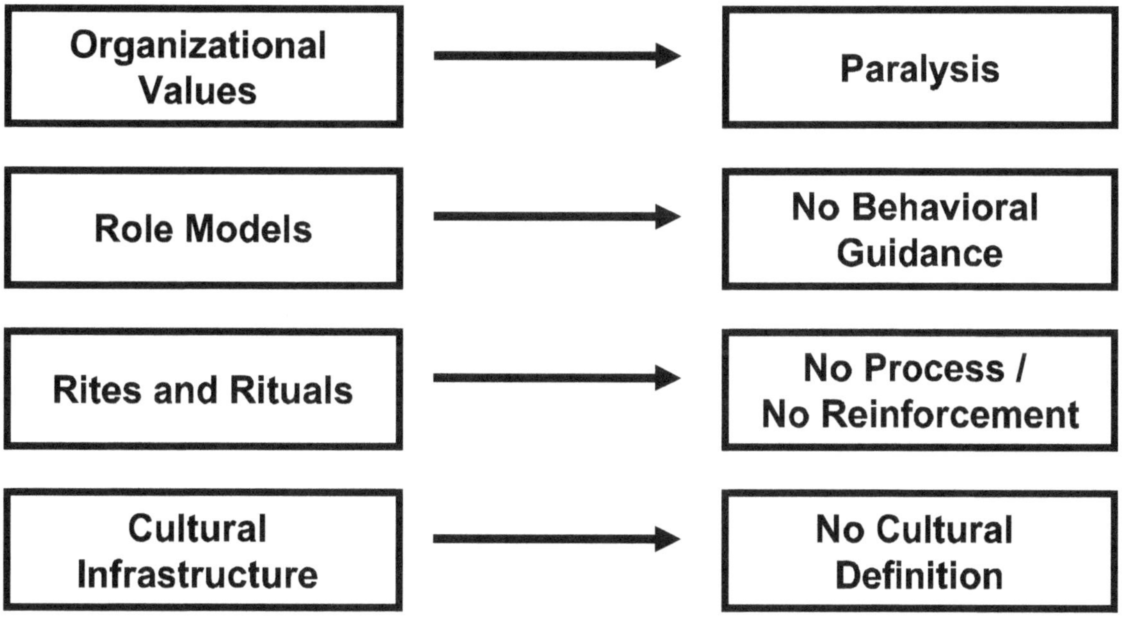

Exercise 8-3 The Impact of Missing the Elements of Culture Part 1

Chapters 4 through 7 discussed the four elements of culture. When properly in place, they go a long way to supporting the status quo or, if modified correctly, to supporting and reinforcing change initiatives.

After Mike Kane departed, ATPCo's eastern plant faced a period of high stress among those who had been successful within Mike's system of managing the plant. The stress increased when the new manager introduced an entirely different set of organizational values focused on equipment reliability, not rapid repair.

Of one thing everyone was certain: the new manager was approaching the plant's change of focus in the wrong manner. He believed that his authority would allow him to dictate change and, as a result, change would take place simply because he mandated it. New values were issued to the workforce by memorandum. Existing role models were left in place and told that they either changed or would be fired. The alteration to the rites and rituals appeared in ATPCo's new work process manual with little or no training. Those within the cultural infrastructure continued as if Mike was still the manager.

Where do you think the manager's problems lie? For each of the four elements of culture, explain what was missed and what you would do differently.

Element	What Was Missed	What Would You Do Differently?
Organizational Values		
Role Models		
Rites and Rituals		
Cultural Infrastructure		

Exercise 8-3 The Impact of Missing The Elements of Culture Part 2

Try the same exercise for a change that has taken place in your own plant. Think about the change process, then identify what was missed related to the four elements of culture.

Element	What Was Missed	What Would You Do Differently?
Organizational Values		
Role Models		
Rites and Rituals		
Cultural Infrastructure		

Figure 8-7 The Relationship Between the Elements of Change and Culture

	Values	Role Models	Rites & Rituals	Cultural Infrastructure
Leadership	M	M	M	m
Work Process	M	M	M	m
Structure	M	m	M	m
Group Learning	M	M	m	m
Technology	M	M	M	m
Communication	M	M	M	M
Interrelationships	M	M	m	M
Rewards	M	M	m	m

M = Major interaction m = minor interaction

Figure 8-8 How Do the Elements of Change and Culture Impact Each Other?

What are the issues within the element of

(select one of the four elements of culture)

that impacts

(select one of the eight elements of change)

What needs to be changed
to bring them into alignment
with what is being implemented?

Exercise 8-4 The Impact of the Elements on Each Other

It is important to recognize that the four elements of culture and the eight elements of change work hand-in-hand. As you develop plans for deploying your change initiative, you need to take this interrelationship into consideration. The elements of culture impact the elements of change and vise versa. This exercise tests your understanding of this fact. Think about an initiative in which you have been involved. Ask yourself the question shown in Figure 8-8. Then write this information on the following chart, at the intersection of each element of change and culture.

Suppose an initiative called for work planning and scheduling practices to be followed in order to optimize the use of the maintenance resources. You would ask yourself, – "For the planning and scheduling initiative, what are the issues with organizational values (or one of the other elements of culture) that impacts leadership (or one of the other eight elements of change)?" What needs to be changed to bring them into alignment with what is being implemented?

The answer to this specific question would be that the organizational values associated with a planning and scheduling process must be championed by the leadership in order to have the initiative be successful. If this is not the case, then the values need to be altered. The same question relating role models with leadership would indicate that it is incumbent upon the leadership to have the correct role models in place to support the planning and scheduling process.

	Organizational Values	Role Models	Rites and Rituals	Cultural Infrastructure
Leadership				
Work Process				
Structure				
Group Learning				
Technology				
Communication				
Interrelatiuonships				
Rewards				

Leadership

Chapter 9

Purpose

This chapter introduces the first and most important of the eight elements of change—leadership. You can struggle to implement a change initiative without some of the other elements, but without leadership, all is lost.

What Is Included

Figure 9-1 Leadership Defined

Figure 9-2 Desire vs. Skills

Exercise 9-1 Desire vs. Skills

Figure 9-3 Relationship vs. Task

Exercise 9-2 Your Relationship vs. Task Profile

Figure 9-4 Transactional vs. Transitional Skills

Exercise 9-3 Your Transactional / Transitional Profile

Figure 9-5 Power Defined

Figure 9-6 Types of Power

Exercise 9-4 Recognizing Types of Power

Exercise 9-5 Leadership and Culture

Exercise 9-6 The Leadership Web

Exercise 9-7 The Leadership Web – Analyzing the Low Spots

Figure 9-1 Leadership Defined

A change-management leader
is a person who can clearly see
and communicate a vision of the future,
influence others to embrace the changes
necessary to accomplish that vision,
maintain organizational focus over an extended
time period, and support group learning
as the process evolves.
It boils down to "what we conceive
and believe, we can achieve."
The leader is the person who can pull it all together.

Figure 9-2 Desire vs. Skills

	They do not have the desire	They have the desire
They have the Skills	Quadrant Three They have the skill but not the desire. Coaching or more severe measures may be needed to get these people "on board."	Quadrant Four Ideal situation. These are the people who will lead the change process and help to provide the related improvements.
They do not have the skills	Quadrant One The worst of all possible situations. These people are probably not going to help the change, or any other effort.	Quadrant Two They have the desire. That's good. Training or experience may be all that is needed.

Exercise 9-1 Desire vs. Skills

Leaders can fit any of the quadrants identified in Figure 9-2. This exercise helps you identify the quadrant in which each of the leaders in each scenarios fits, and then to suggest what you, as their manager, would do to help them reach Quadrant Four.

Scenario	Quad	Corrective Action
John, the lead foreman of the machine shop, has worked as a mechanic as well as a foreman and is highly respected for his work. The problem is he arrives late, leaves early, and is often hard to find when things need to be done.		
Peter, one of the foremen that you just promoted, is always trying to do the right things to get the job done. However, he lacks the field experience and often flounders.		
As a manager, your biggest problem is Sue, the lead accountant. In her position, she makes mistakes and becomes defensive when confronted with the results of these errors. The thing she talks about the most is her upcoming retirement.		
Arthur was just promoted to a managerial position at another plant. His high level of energy and his skills were highly respected by his crew as well as his peers.		

Scenario	Quad	Corrective Action
Scott is a highly-skilled computer technician who you just recently promoted to group leader. On more than one occasion, his clients have told you that he doesn't seem to care about the work.		
Kate, your most recently appointed lead engineer, worked very hard to get her new job. During a conversation, she asked if the company would pay for her master's degree so that she could learn additional aspects of the job with which she was not familiar.		
When you asked your lead inspector to take on an important assignment, it was clear from his attitude that he really wasn't interested. Later, you found out that he had called in sick to avoid the training needed for the task you offered him.		
Allie, the manager of your internet communication department, comes in early and stays late. Her focus is on the job, not the clock. Her work is always of the highest quality. Additionally she just received an advanced degree qualifying her for more challenging assignments.		

Figure 9-3 Relationship vs. Task

	Low Task	High Task
High Relationship	Quadrant Three People skills are high which could serve to motivate the organization. However, task focusing skills are low. May not stay the course.	Quadrant Four Ideal situation. The leader has good skills in both the relationship and task areas.
Low Relationship	Quadrant One The worst of all possible sitiations. The leader does not have skills in either of the critical areas.	Quadrant Two Task skills are high so there can be focus on the work. However, people skills are low and there may be motivational or credibility issues.

Exercise 9-2 Your Relationship vs. Task Profile
Part 1

For each question, circle the answer that indicates your feeling about the statement as it applies to you: SA= strongly agree, A=agree, N= no opinion, D= disagree and SD = strongly disagree. Total the number of responses in each column. Notice that the order of responses changes from row to row. Answer carefully.

1.	I have a good working relationship with managers, peers, and subordinates.	SD	D	N	A	SA
2.	As a leader, I stress task completion and exp ect others to follow my lead.	SA	A	N	D	SD
3.	When I work, I get very focused on the task. I have no time for socializing.	SA	A	N	D	SD
4.	I like working in groups and being part of what they accomplish.	SD	D	N	A	SA
5.	I feel very comfortable leading a task and having to engage others in the effort.	SD	D	N	A	SA
6.	When it comes to accomplishing a task, people are the most important ingredient.	SD	D	N	A	SA
7.	I would rather work alone.	SA	A	N	D	SD
8.	In meetings, I listen carefully to what others have to say and take their opinions into consideration.	SD	D	N	A	SA
9.	I have problems leading or even working in groups. They take too long to get the work accomplished. "Just tell me what to do and let me do it."	SA	A	N	D	SD
10.	In meetings, I focus on the task content. The discussions often seem to be a waste of my time.	SA	A	N	D	SD
Total the number of responses in each column. Write the number in the boxes at right.						

Exercise 9-2 Your Relationship vs. Task Profile Part 2

Suppose the sum of each column from Part 1 was:

2	5	1	2	0

The plot for this information follows:

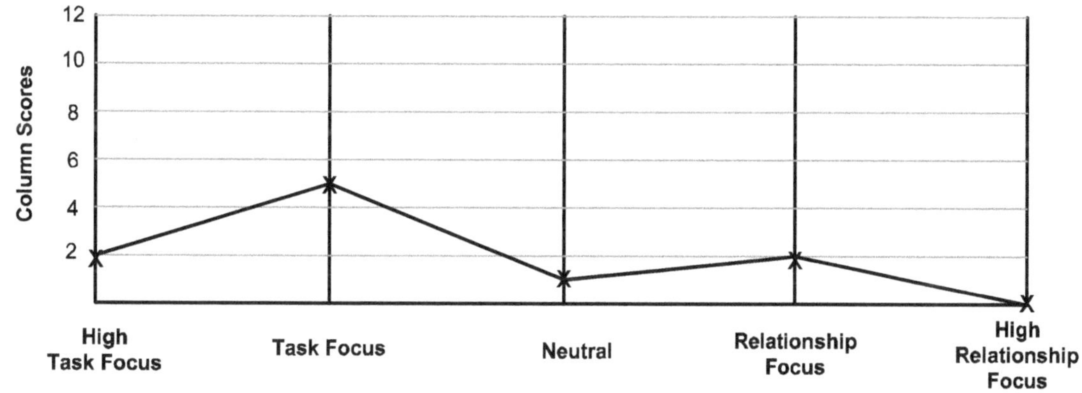

On the chart below, mark an X indicating how many times you selected that column in Part 1 of this exercise. Then draw a line connecting the Xs.

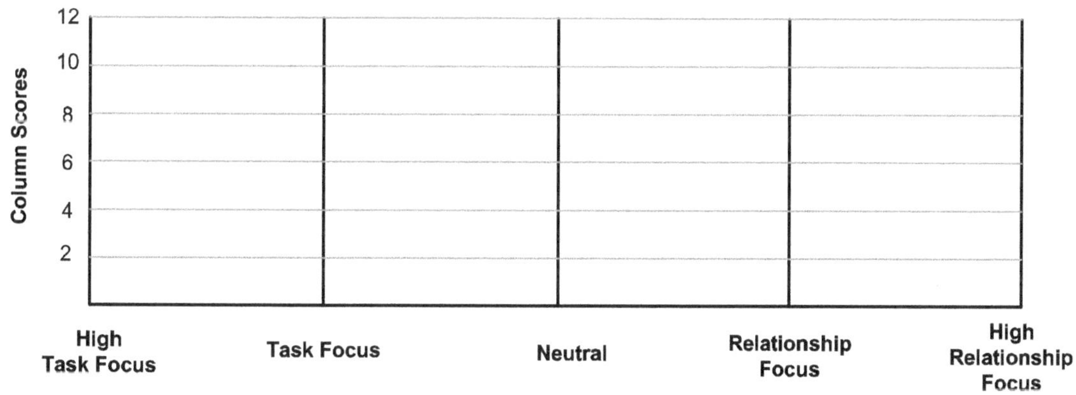

What does your survey tell you about your relationship / task profile?

Figure 9-4 Transactional vs. Transitional Skills

	Low Transitional Skills	High Transitional Skills
High Management - Transactional Skills	Quadrant Three Leaders have a higher level of transactional skills. They can do the work, but leading into a new order of things is less likely.	Quadrant Four Ideal situation. The leader has good skills in both areas.
Low Management - Transactional Skills	Quadrant One The worst of all possible situations, the leader does not have skills in either of the critical areas.	Quadrant Two Leaders have a higher degree of transitional skills. These people can conceptualize the change, but have trouble making it happen.

Exercise 9-3 Your Transactional / Transitional Profile Part 1

For each question, circle the answer that indicates your feeling about the statement as it applies to you: SA = strongly agree, A = agree, N = no opinion, D = disagree, and SD = strongly disagree. Total the number of responses in each column. Notice that the order of responses changes from row to row. Answer carefully.

1.	I like to work on projects where I get to develop detailed solutions.	SA	A	N	D	SD
2.	I like to create projects, and then have others develop and work through the details.	SD	D	N	A	SA
3.	I like to solve problems associated with the detailed aspects of the work.	SA	A	N	D	SD
4.	I like to attend conferences, acquire new information, and adapt it to the business.	SD	D	N	A	SA
5.	People see me as the resident expert in one or more aspects of the work.	SA	A	N	D	SD
6.	I like to read books and periodicals, and to look for business opportunities.	SD	D	N	A	SA
7.	I like to complete detailed plans associated with getting things done.	SA	A	N	D	SD
8.	I invent new goals and initiatives for the company to develop.	SD	D	N	A	SA
9.	I like working at the goal–initiative level of the Goal Achievement Model.	SD	D	N	A	SA
10.	I like working on the activities and measures of the Goal Achievement Model.	SD	D	N	A	SA
Total the number of responses in each column. Write the number in the boxes at right.						

Exercise 9-3 Your Transactional / Transitional Profile

Suppose the sum of each column from Part 1 was:

2	5	1	2	0

The plot for this information follows:

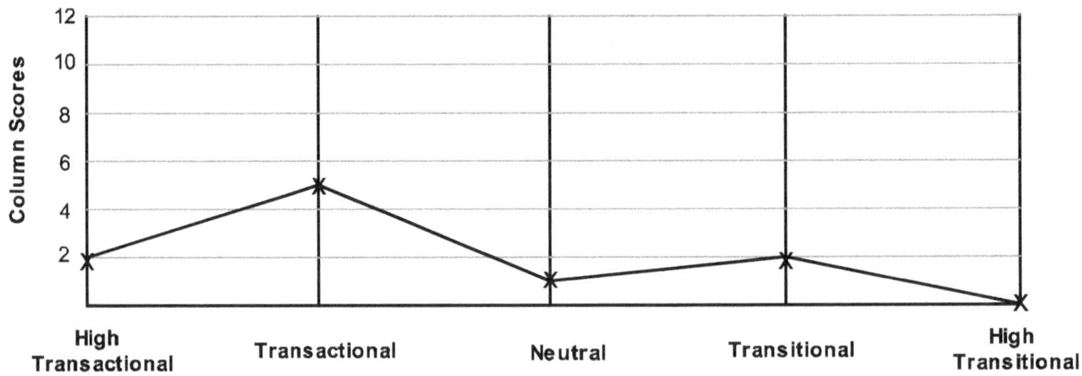

On the chart below, mark an X indicating how many times you selected that column in Part 1 of this exercise. Then draw a line connecting the Xs.

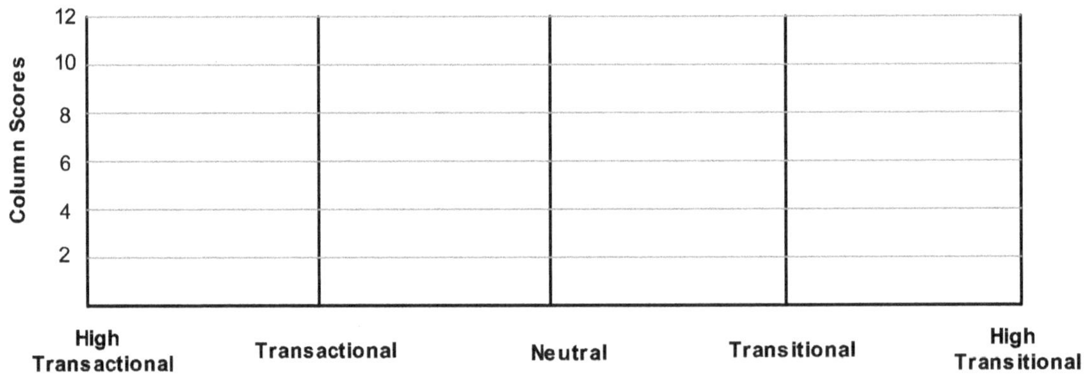

What does your survey tell you about the type of transactional or transitional roll you fill?

Figure 9-5 Power Defined

Power is "the capacity to affect the behavior of others in a desired direction." There are different types of power, some good and some not. Each type, and the way each is used, affects people's behavior differently. Thus, power and its use directly impact the relationship between leaders and their groups. The change process is more likely to be successful when the relationship is positive than when the relationship is poor.

Figure 9-6 Types of Power

Type of Power	Typical Organization Reaction
Legitimate Power Provided to the person by the organization	**Compliance** at least on an overt level. How the person acts and uses the power bestowed will determine if the compliance turns to commitment or resistance.
Reward Power Based on the ability to provide rewards	**Compliance** in order to acquire the reward, assuming that the reward is something desired.
Coercive Power Ability to punish for inaction	**Resistance** is usually the outcome of application of this form of power. People want to be treated with dignity and respect.
Expert Power Based on possession of information	**Commitment** so long as this form of power is not abused. For change to succeed, information needs to be shared.
Information Power Access to and control over information	**Compliance** which can turn into resistance if the needed information is withheld.
Referent Power Leader is admired, identified with	**Commitment** because these individuals are usually natural leaders.

Exercise 9-4 Recognizing Types of Power
Part 1

Recognizing the types of power used by various leaders within the organization is important. Each type of power elicits an organizational response. Therefore, you must use those types of power that help the organization change for the better, and avoid the types of power that can inhibit the process.

Scenario	Type of Power and Potential Outcome
Albert, the manager of the systems group, is always sought out for his opinion when critical decisions are needed.	
Betty, the production area manager, always gets her people to follow her directives. Those who don't wind up with less-than-desirable work assignments.	
Frank, the plant manager, is highly respected for his technical and people skills. The plant is committed to helping him achieve the plant goals.	
June, the technical manager, has an excellent grasp of the plant process and how to improve it. However, she doesn't share this information, expecting everyone to come to her when they need help with a process problem.	
Joe, the maintenance manager, gets compliance by using the power associated with his position. He has very good people skills, so compliance is easy to attain.	
Everyone at ATPCo knew that if you didn't do what Mike Kane directed, you would not get a decent salary increase at your yearly anniversary. You could also be fired.	

Exercise 9-4 Recognizing Types of Power in Your Organization Part 2

Select two or three leaders within your own organization. Then answer the questions related to the power they exert.

Selection: (position in the organization)
Over whom do they have power?
What type of power do they exert?
What is the organization's reaction to the power?
If it is positive, why do you think it has reacted this way?
If it is negative, why do you think it has reacted this way?
How do you think the negative power issue could be corrected?

Exercise 9-5 Leadership and Culture
Part 1

With the introduction of Epsilon, ATPCo's eastern plant has a serious problem. Demand was high, but the reliability of their equipment was not. With his "run it hard, fix it fast" philosophy, Mike Kane had run the equipment into the ground. Now when the equipment was needed the most, it often was not available. The production manager of the Epsilon lines, Pat Hawkins, held this same position during Mike Kane's tenure and did very well as he applied Mike's strategy to his units. However, the new manager is much more focused on reliability of the equipment, realizing that, without this focus, he will never be able to supply the market.

Pat talks about reliability and promotes it at his staff meetings. However, he is much more interested and involved with the work when things break down. At these times, he is out on the production line, focused on getting the equipment back into service. The recently-hired reliability engineer for the Epsilon line has been very frustrated during these events because Pat ignores him and has no time for his recommendations about reliability improvements. In fact, last week when one of the critical pumps failed, Pat got it back into service so quickly that the engineer never even had time to tell Pat that he had an improved seal that would extend the run life dramatically.

As Pat's manager, you see other problems as well. Pat has been slow to hire new engineers to help him with his reliability problems. He has a budget for three engineers and has only hired one. You know that he has had resumes for others with very strong skills on his desk.

The working relationship he has with production shift foremen and with his maintenance foreman seem to be excellent, but when you look closer you find that they are just as focused on the "run it hard, fix it fast" philosophy as is Pat. Had you not gone into the plant and talked with the operators and maintenance mechanics, this would not have been known because all of the supervisors preach reliability at your team meetings.

The other thing you discovered when you talked with the line operators and others in the plant was that Pat spends a considerable amount of time trying to discover reliability initiatives that did not work. He then spreads the word across the plant through his numerous contacts, telling them how the old way worked well and this "new reliability initiative" is a waste of time.

Questions:

How has Pat affected the organizational values of the plant?
How are the new rituals and their reinforcing rites affected by his behavior?
How well is he as a representative role model of the new reliability culture?
What role does he play in the eastern plant's cultural infrastructure?
If you were Pat's supervisor, what would you do?

Exercise 9-5 Leadership and Culture Part 2

Select one of you own leaders and answer the same questions from Part 1.

Questions:

How has the leader you selected affected the organizational values of the plant?
How are the new rituals and their reinforcing rites affected by the leader's behavior?
How well are they a representative role model for the culture?
What role do they play in the plant's cultural infrastructure?
What are the positive impacts they have made on the culture?
What are the negatives impacts?
If you were the plant manager, what would you do to correct the negatives?

Exercise 9-6 The Leadership Web

The leadership section of the web of change survey is divided into four parts, as shown below. By answering the survey's 16 questions and entering your scores on the four-axis web, you will see how leadership scores within the four elements of culture.

Scores:

Category	Questions	Score
Organizational Values	1 to 4	
Role Models	5 to 8	
Rites and Rituals	9 to 12	
Cultural Infrastructure	13 to 16	
Total Score – Leadership	———	

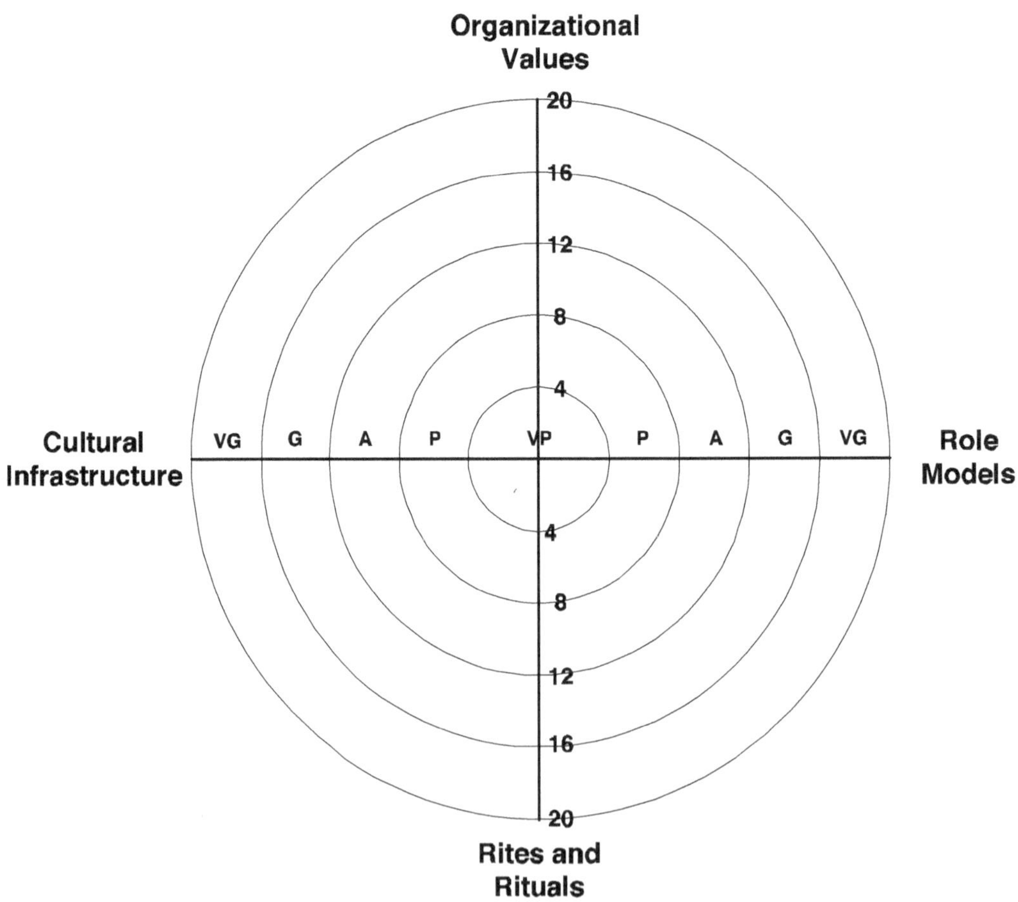

A.	**Leadership**	**Points**
1	Leadership has established a reliability-focused value system.	
	A. Strongly agree	5
	B. Agree	4
	C. Neutral	3
	D. Disagree	2
	E. Strongly disagree	1
2	Leadership has communicated this value system throughout the plant.	
	A. Strongly agree	5
	B. Agree	4
	C. Neutral	3
	D. Disagree	2
	E. Strongly disagree	1
3	The value system at the site has been audited to assure that it is understood at all levels of the organization.	
	A. Strongly agree	5
	B. Agree	4
	C. Neutral	3
	D. Disagree	2
	E. Strongly disagree	1
4	The site leadership demonstrates through their behavior and actions that they believe and are following the established values.	
	A. Strongly agree	5
	B. Agree	4
	C. Neutral	3
	D. Disagree	2
	E. Strongly disagree	1
5	Role models within the organization are recognized by those in leadership positions.	
	A. Strongly agree	5
	B. Agree	4
	C. Neutral	3
	D. Disagree	2
	E. Strongly disagree	1

6 The site role models support the change and new ways for conducting the work.

A. Strongly agree 5

B. Agree 4

C. Neutral 3

D. Disagree 2

E. Strongly disagree 1

7 Leadership clearly understands the impact that role models have on the work process, specifically a process that is undergoing change.

A. Strongly agree 5

B. Agree 4

C. Neutral 3

D. Disagree 2

E. Strongly disagree 1

8 Role models who do not support the change effort are reassigned or removed from their area of influence.

A. Strongly agree 5

B. Agree 4

C. Neutral 3

D. Disagree 2

E. Strongly disagree 1

9 The site's rituals and their supporting rites are clearly understood.

A. Strongly agree 5

B. Agree 4

C. Neutral 3

D. Disagree 2

E. Strongly disagree 1

10 The rites and rituals that are out of sync with the new process are being altered.

A. Strongly agree 5

B. Agree 4

C. Neutral 3

D. Disagree 2

E. Strongly disagree 1

11	Old rites and rituals are not being reinforced.	
	A. Strongly agree	5
	B. Agree	4
	C. Neutral	3
	D. Disagree	2
	E. Strongly disagree	1
12	New rites and rituals are being reinforced by the site leadership and others.	
	A. Strongly agree	5
	B. Agree	4
	C. Neutral	3
	D. Disagree	2
	E. Strongly disagree	1
13	Leadership understands the importance of addressing the cultural infrastructure as part of any change effort.	
	A. Strongly agree	5
	B. Agree	4
	C. Neutral	3
	D. Disagree	2
	E. Strongly disagree	1
14	The Keepers of the Faith have been involved with the development of the change initiative.	
	A. Strongly agree	5
	B. Agree	4
	C. Neutral	3
	D. Disagree	2
	E. Strongly disagree	1

15 Gossips, whisperers, and spies are identified and have been part of the process to some extent.

A. Strongly agree	5
B. Agree	4
C. Neutral	3
D. Disagree	2
E. Strongly disagree	1

16 Language and symbols have been developed to support the change effort.

A. Strongly agree	5
B. Agree	4
C. Neutral	3
D. Disagree	2
E. Strongly disagree	1

Exercise 9-7 The Leadership Web – Analyzing the Low Spots

This exercise will help you use the leadership web survey that you completed in Exercise 9-6 to address potentially-needed corrective actions within the four elements of culture.

Enter Your Leadership Survey Scores.

Scores:

Category	Questions	Score	VP 1-4	P 5-8	N 9-12	G 13-16	VG 17-20
Organizational Values	1 to 4						
Role Models	5 to 8						
Rites & Rituals	9 to 12						
Cultural Infrastructure	13 to 16						

Enter your scores from the leadership web survey in the "Score" column. Then mark an X in the appropriate column based on the following key. Scores are classified as: VP = very poor (1 – 4 points); P = poor (5 – 8 points); A = average (9 – 12 points); G = good (13 – 16 points); and VG = very good (17 – 20 points).

2. Questions:

Which cultural scores are low (points between 1 and 8)?
Why do you believe that these cultural areas have scored this low?
From a leadership perspective, what could be done to make positive improvement?
Which cultural scores are high (points between 13 and 20)?
Why do you believe that these cultural areas have scored this high?
From a leadership perspective, what could be done to generate improvements in the low scoring areas?

Work Process

Chapter 10

Purpose

This chapter introduces the second of the eight elements of change—work process. This element addresses the process by which work gets accomplished as well as who has responsibility for its performance.

What Is Included

Figure 10-1	Work Transforms
Figure 10-2	Workflow / Information Flow Diagram
Figure 10-3	Steps to Change the Work Process
Exercise 10-1	Changing the Work Process – What is Missing?
Figure 10-4	Approaches to Process Change
Exercise 10-2	Which Approach Would You Choose?
Figure 10-5	Simple Workflow
Figure 10-6	Complex Workflow
Figure 10-7	RACI Charting
Figure 10-8	RACI Guidelines
Figure 10-9	The RACI Identifiers
Figure 10-10	The RACI Format
Exercise 10-3	RACI Practice
Exercise 10-4	Work Process Web
Exercise 10-5	The Work Process Web – Analyzing the Low Spots

Figure 10-1 Work Transforms

Simple

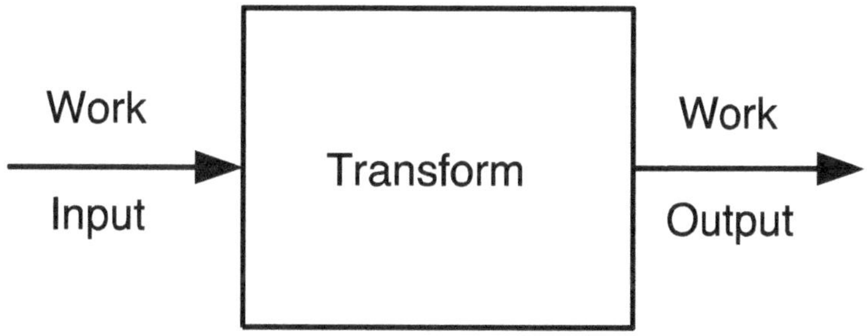

Complex

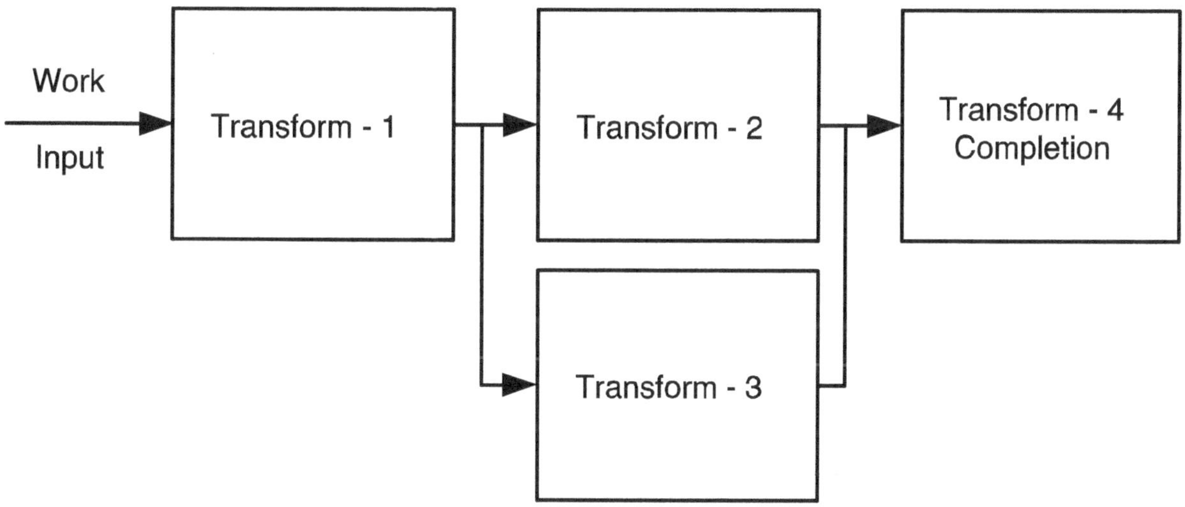

Figure 10-2 Workflow / Information Flow Diagram

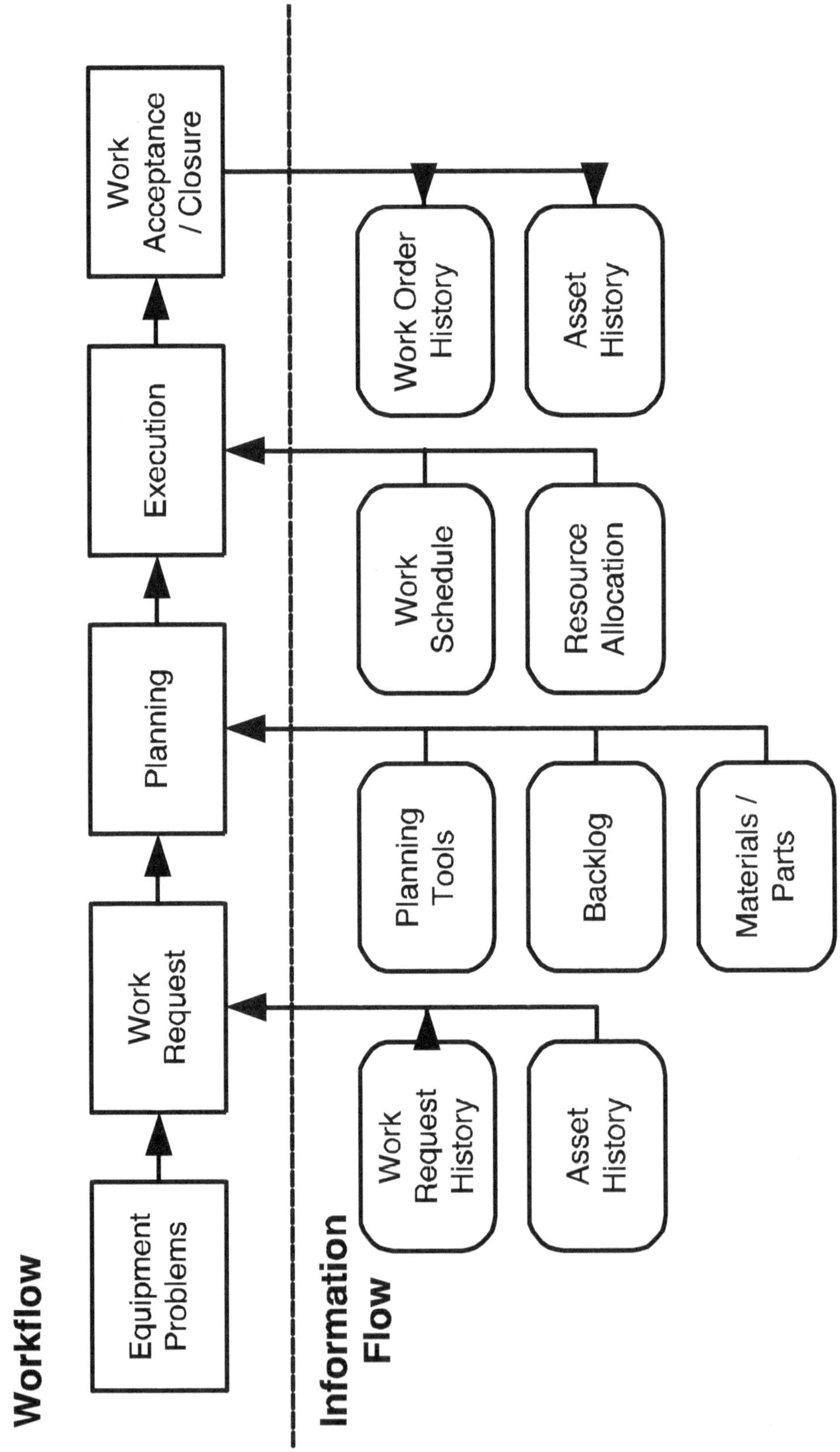

Figure 10-3 Steps to Change the Work Process

1. Recognize that change is needed.

2. Determine who needs to be involved in building the new process.

3. Be clear on the values you are trying to achieve.

4. Use the Goal Achievement Model.

5. Understand the current set of rituals and reinforcing rites.

6. Conduct the work process design meetings.

7. Validate the work.

8 Communicate.

9. Training

10. Care for and feed the process (sustainability).

Exercise 10-1 Changing the Work Process – What is Missing?

In Figure 10-3, I described ten steps needed to successfully change the work process. Ten scenarios are listed below; in each case, one of the steps was missed. Identify the step that was missed as well as what you would do to address the issue.

Scenario	Missed Step#	What Would You Do?
The work process was redesigned, but the redesign team never tested their ideas with anyone.		
The senior management wanted a reliability-focused design, but the organization believed that their responsibility was to repair what broke down as quickly as possible and return it to service.		
Everyone believed that the status quo has worked well for years and there was no need to change.		
When asked about goals, initiatives, or activities to support the change, management had nothing prepared.		

Scenario	Missed Step#	What Would You Do?
The redesign was done by the maintenance and production managers without any input from the rest of the organization.		
A manual describing the process was distributed; everyone was told that by reading it they would be prepared for the work process change that was being instituted.		
The maintenance manager believed that he and the production manager were the only people who needed to be involved in the redesign because they knew all that was required.		
No one in the organization knew what was being changed or why. When the redesign was deployed, everyone was surprised.		
The new work process ran into problems as soon as it was rolled out. People found that there was conflict between what they were doing as part of their day-to-day activities and what the new process expected of them.		

Figure 10-4 Approaches to Process Change

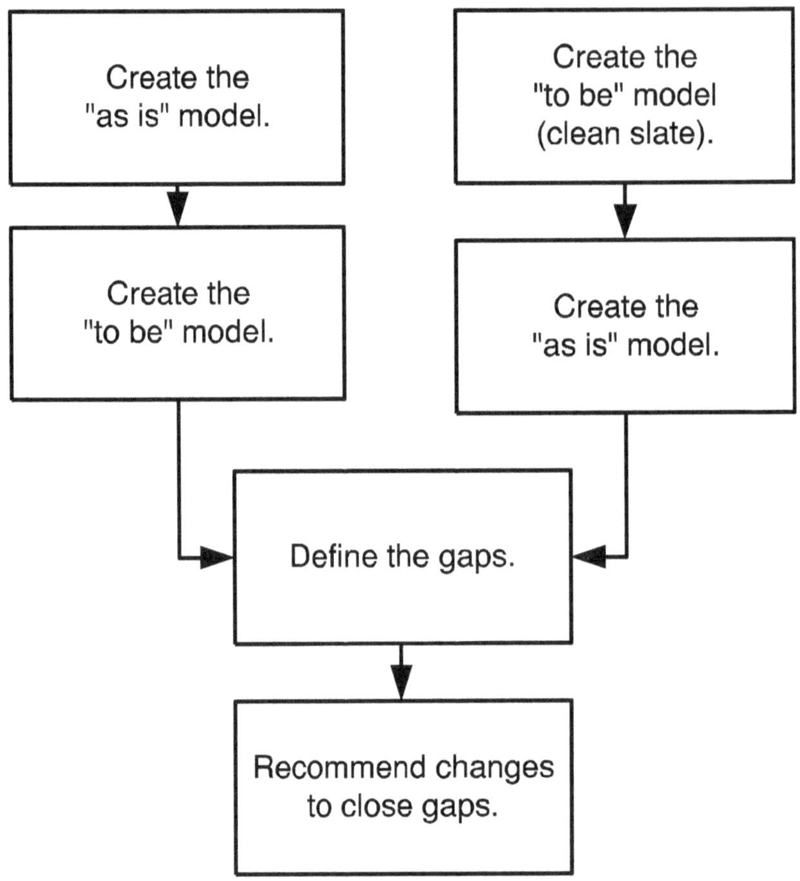

Exercise 10-2 Which Approach Would You Choose?

For the following examples, choose which approach you would use: starting with the "as is" model or with the clean slate model. Briefly explain your choice in the appropriate block.

Example	As Is	Clean Slate	Explain Why.
1 ATPCo changes from a functional organization to one developed around teams tied to the production lines.			
2. A plant changes its work process from one that repairs things that break to one that is focused on preventing failure.			
3. A work management process is not working and is costing the firm money. The process is in need of radical overhaul.			
4. The purchasing work process is being consolidated (fewer job steps and people). The current process won't function in the new work environment.			
5. The machine shop needs to upgrade its work process in support of the new highly-technical equipment being installed.			
6. A new manager takes over the maintenance department. He feels that the old process is outdated and needs to be changed.			

Figure 10-5 Simple Workflow

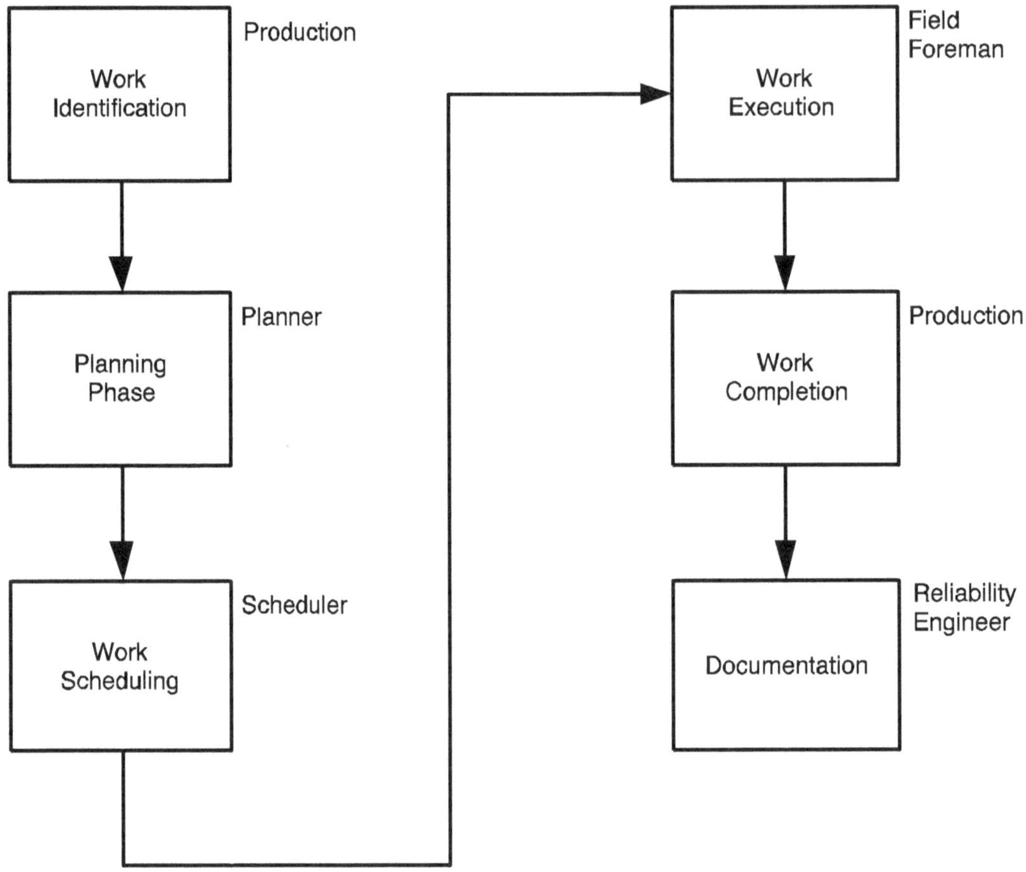

Figure 10-6 Complex Workflow

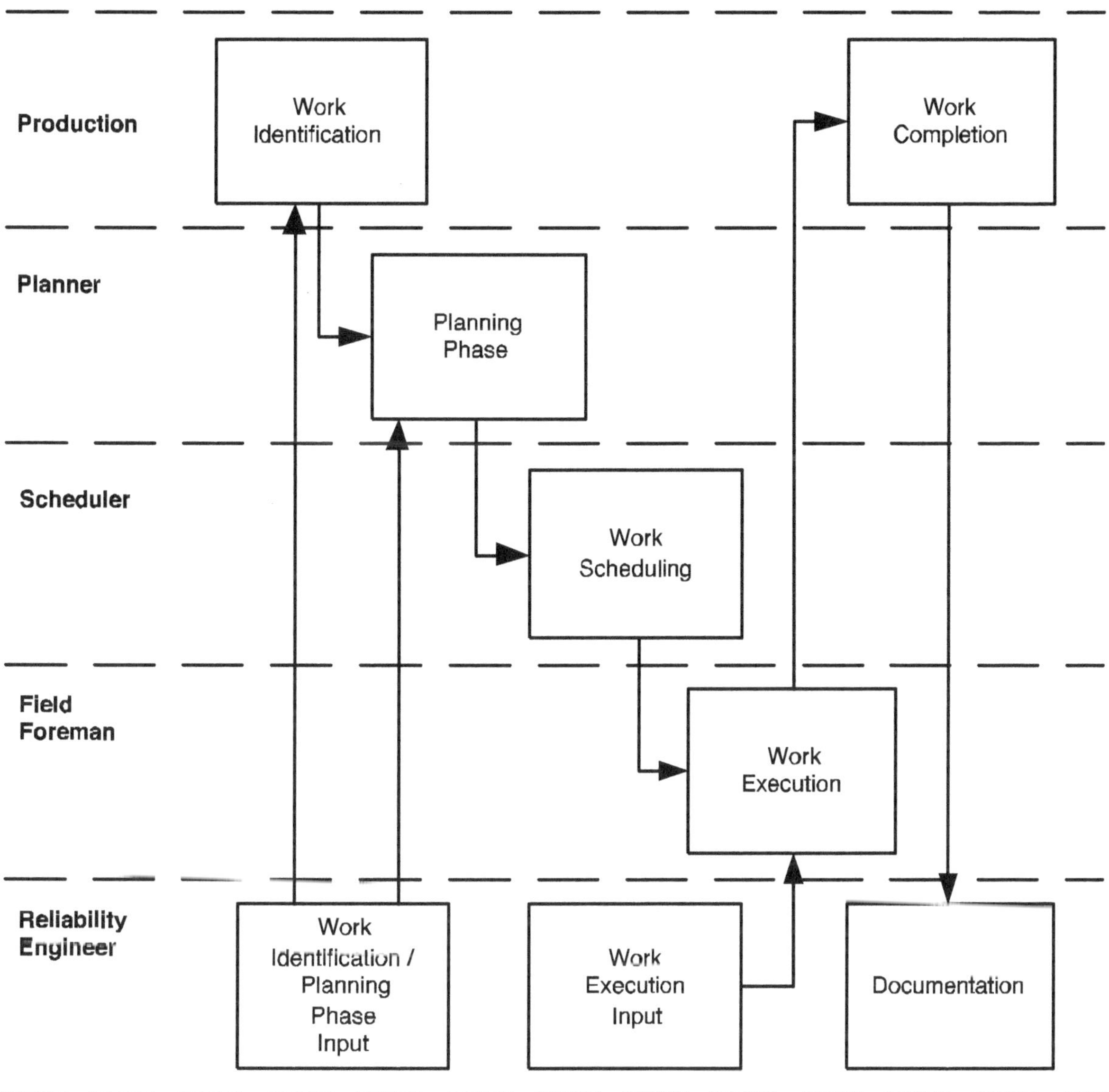

Figure 10-7 RACI Charting

A technique for identifying key work activities and functional areas of responsibility associated with a specific task in a matrix format. By representing the work task activities in this fashion, the roles and responsibilities of the functions associated with each task can be clearly delineated.

Figure 10-8 RACI Guidelines

1. Conduct the analysis with a team of experts.
2. Take time to do it right.
3. Place accountability (A) and responsibility (R) where they truly belong.
4. There is only one "A" per activity – the person solely accountable for the task. An "A" can also have "R" functions.
5. There can be multiple "R" roles for each task; however, too many Rs can cause problems.
6. Authority and accountability for action MUST be linked.
7. Only assign consult (C) and inform (I) roles with great care. Not every task requires them.

Figure 10-9 The RACI Identifiers

Accountable	The position with sole (go / no go) authority over the task to be completed.
Responsible	These are the people who are working on the task. Responsibility can be shared as determined by the "A" person.
Consult	These people are consulted prior to the final decision or action. This requires two-way communications.
Inform	These people are informed about the decision or action. This does not have to be prior to the decision being made.

Figure 10-10 The RACI Format

	Functions of the various groups involved with the tasks					
Tasks associated with the activity under consideration						
↓						

To conduct a RACI review, you need to analyze the activity that is being performed and each of the tasks that comprise the activity. In the left column, list each task that is a part of the activity being analyzed. Across the top row, list each functional area involved with the task. Then in the intersecting cells, identify the role and responsibility of the function (A, R, I, C, or blank) for that task. See text pages 158-159 for an example.

Exercise 10-3 RACI Practice

The manufacturing of tractium products, as with any production process, results in equipment deterioration to the eventual point where components fail to perform per design. As a result, production suffers. When this situation occurs, production submits a work order to the maintenance planning group. In turn, they develop a detailed work plan for execution of the repair. Included in this plan is input from the reliability engineers who analyze the problem and add work steps that will improve the reliability of the equipment, once it is returned to service. Also included is materialization of the job so that when the repair is in progress, the parts are available for the work.

When the work plan is completed, it is referred to as “ready to schedule.” At this point, production determines when it will be scheduled and ultimately worked on by maintenance. When the job is completed, production is informed that the work is finished. They inspect and approve the job, at which point planning is alerted that the work order can be closed.

Using the following workflow diagram, develop a RACI chart for this series of work activities

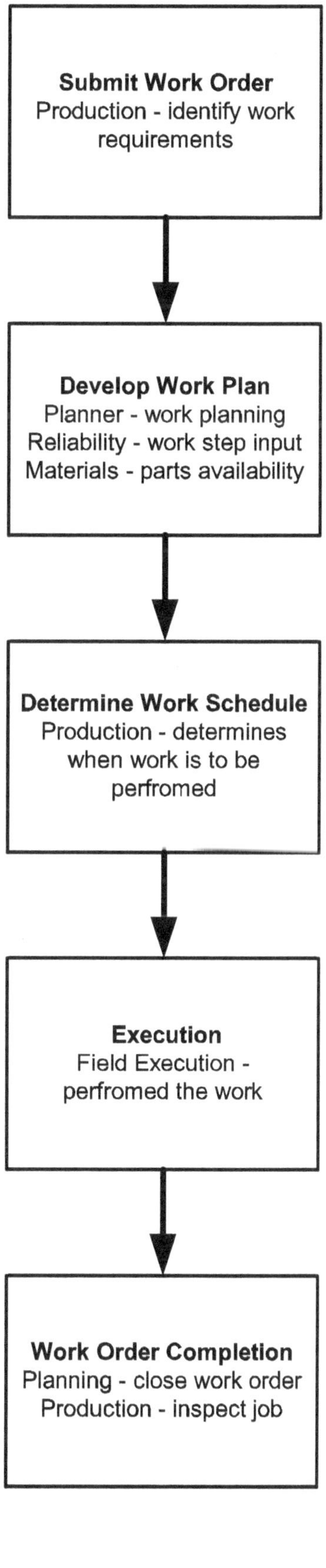

Exercise 10-3 RACI Form

	Functions					
Tasks						

Exercise 10-4 Work Process Web

The work process section of the web of change survey is divided into four parts, as shown below. By answering the survey's 16 questions, and entering your scores on the four-axis web, you will see how work process scores within the four elements of culture.

Scores:

Category	Questions	Score
Organizational Values	1 to 4	
Role Models	5 to 8	
Rites and Rituals	9 to 12	
Cultural Infrastructure	13 to 16	
Total Score – Work Process	———	

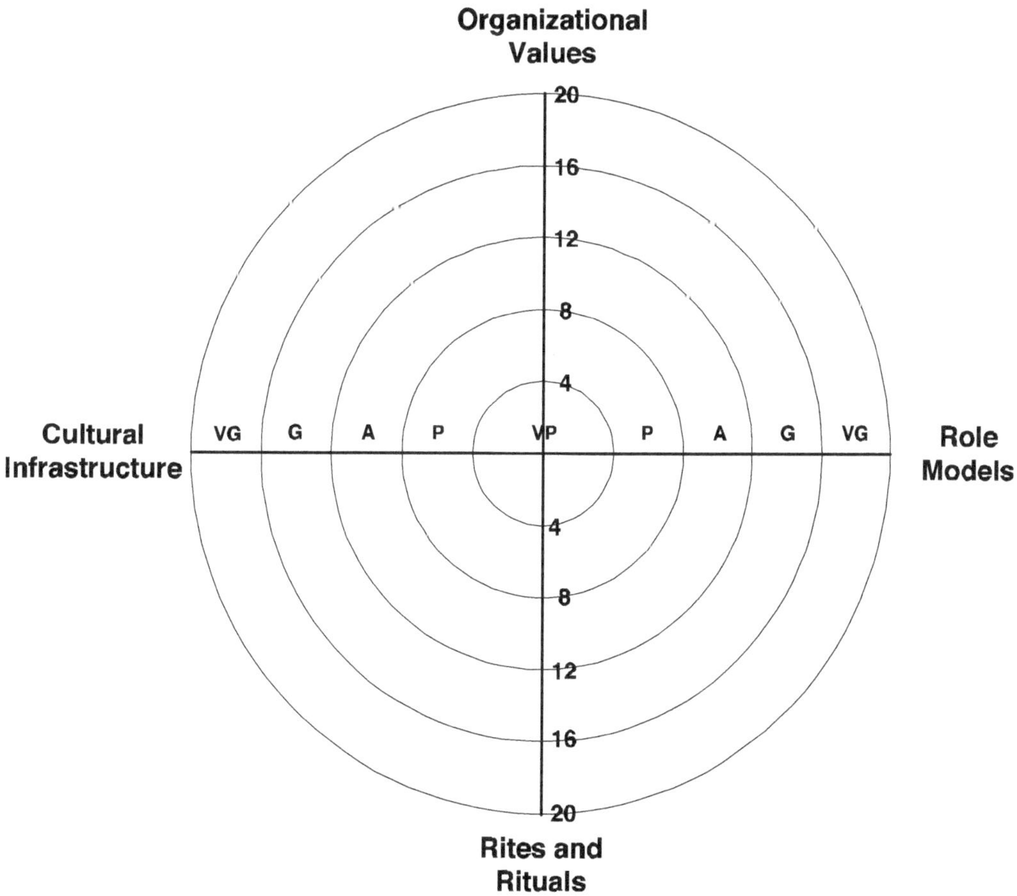

B. Work Process

		Points
1	The work process has been designed to support the organizational values.	
	A. Strongly agree	5
	B. Agree	4
	C. Neutral	3
	D. Disagree	2
	E. Strongly disagree	1
2	The work process is periodically audited to assure alignment, detect issues, and address corrective action.	
	A. Strongly agree	5
	B. Agree	4
	C. Neutral	3
	D. Disagree	2
	E. Strongly disagree	1
3	The work process includes the ability to accurately capture information about the plant assets in such a way that the information can be utilized in support of the process and related values.	
	A. Strongly agree	5
	B. Agree	4
	C. Neutral	3
	D. Disagree	2
	E. Strongly disagree	1
4	The work process has clearly delineated how the work flow and information flow models align with each other.	
	A. Strongly agree	5
	B. Agree	4
	C. Neutral	3
	D. Disagree	2
	E. Strongly disagree	1

5	The role models have actively supported the work process design efforts that focus the organization on improved plant reliability.	
	A. Strongly agree	5
	B. Agree	4
	C. Neutral	3
	D. Disagree	2
	E. Strongly disagree	1
6	The role models have actively supported new and improved work processes when they are rolled out and implemented within the plant.	
	A. Strongly agree	5
	B. Agree	4
	C. Neutral	3
	D. Disagree	2
	E. Strongly disagree	1
7	The role model's behavior follows the work process as designed and implemented. They "walk the talk."	
	A. Strongly agree	5
	B. Agree	4
	C. Neutral	3
	D. Disagree	2
	E. Strongly disagree	1
8	Role models who are not in support of the work process are identified and corrective actions are taken. These actions could be coaching or other forms.	
	A. Strongly agree	5
	B. Agree	4
	C. Neutral	3
	D. Disagree	2
	E. Strongly disagree	1

9 The rituals that existed in the former work process have been altered or removed by those that support the new process.

A. Strongly agree	5
B. Agree	4
C. Neutral	3
D. Disagree	2
E. Strongly disagree	1

10 The supporting rites (for the rituals referenced in item #9) have been altered or removed to support the new process.

A. Strongly agree	5
B. Agree	4
C. Neutral	3
D. Disagree	2
E. Strongly disagree	1

11 Field review of the process has shown that the former rites and rituals have been removed from the process.

A. Strongly agree	5
B. Agree	4
C. Neutral	3
D. Disagree	2
E. Strongly disagree	1

12 When a rite or ritual that supports the old process (you can't always identify them all at the outset) is identified, corrective action is taken to alter or remove them.

A. Strongly agree	5
B. Agree	4
C. Neutral	3
D. Disagree	2
E. Strongly disagree	1

13 The cultural infrastructure's language and symbols have evolved to incorporate the new process.

A. Strongly agree	5
B. Agree	4
C. Neutral	3
D. Disagree	2
E. Strongly disagree	1

14 The key members of the cultural infrastructure have been identified and their support elicited in developing and rolling out the new process.

A. Strongly agree	5
B. Agree	4
C. Neutral	3
D. Disagree	2
E. Strongly disagree	1

15 A sincere effort has been made to focus stories towards success of the new process vs. how well things were done under the old. The Story Tellers have been used to support his effort.

A. Strongly agree	5
B. Agree	4
C. Neutral	3
D. Disagree	2
E. Strongly disagree	1

16 An effective process is in place to assure that the cultural infrastructure's communication channels (used by the whisperers, gossips, and spies) are focused in support of the new process.

A. Strongly agree	5
B. Agree	4
C. Neutral	3
D. Disagree	2
E. Strongly disagree	1

Exercise 10-5 The Work Process Web - Analyzing the Low Spots

This exercise will help you use the work process web survey that you completed in Exercise 10-4 to address potentially-needed corrective actions within the four elements of culture.

Enter Your Work Process Survey Scores.

Scores:

Category	Questions	Score	VP 1-4	P 5-8	N 9-12	G 13-16	VG 17-20
Organizational Values	1 to 4						
Role Models	5 to 8						
Rites & Rituals	9 to 12						
Cultural Infrastructure	13 to 16						

Enter your scores from the work process web survey in the "Score" column. Then mark an X in the appropriate column, based on the following key. Scores are classified as: VP = very poor (1 – 4 points); P = poor (5 – 8 points); A = average (9 – 12 points); G = good (13 – 16 points); and VG = very good (17 – 20 points)

2. Questions:

Which cultural scores are low (points between 1 and 8)?
Why do you believe that these cultural areas have scored this low?
From a work process perspective, what could be done to make positive improvement?
Which cultural scores are high (points between 13 and 20)?
Why do you believe that these cultural areas have scored this high?
From a work process perspective, what could be done to generate improvements in the low scoring areas?

Structure

Chapter 11

Purpose

This chapter introduces the third element of the eight elements of change—structure. It is the organization's structure that provides the framework upon which a successful change effort can be based.

What Is Included

Figure 11-1 Structure Defined

Figure 11-2 Repair- / Reliability-Based Strategy and Structure

Exercise 11-1 Repair- / Reliability-Based Strategy and Structure

Figure 11-3 Reliability is NOT a Department

Figure 11-4 Process vs. Functional Centric Organizations

Figure 11-5 Process vs. Functional Centric Organizations

Exercise 11-2 Structural Pitfalls

Figure 11-6 Structure Overlaps

Exercise 11-3 ATPCo's Structure

Exercise 11-4 Your Structure

Exercise 11-5 Structural Web of Change

Exercise 11-6 The Structural Web – Analyzing the Low Spots

Figure 11-1 Structure Defined

Structure can be defined as:
a framework of roles, responsibilities,
and reporting relationships that
characterize how a group of
individuals will work together
to achieve a common purpose.

Figure 11-2 Repair- / Reliability-Based Strategy and Structure

A **repair-based strategy** is seen in organizations that have a "fix it now" attitude. The structure is built around the belief that, when equipment breaks, it is maintenance's job to restore it to service as soon as possible

A **reliability-based strategy** is far different. With this kind of strategy in place, the organization is structured to support the belief that things do not fail without warning. The organization also develops reliability-based work plans that proactively take corrective action to prevent failure.

Exercise 11-1 Repair- / Reliability-Based Strategy and Structure

Review the following scenarios and determine whether the organization being described has a repair- or reliability-focused structure and explain why.

Scenario	Repair = REP Reliability = REL	Why
The maintenance department reports in to the production manager and works at his direction on the problem of the day.		
All work is planned with input from the reliability engineers.		
The maintenance department is part of a larger reliability / maintenance team reporting in to the plant manager.		
The maintenance planners and foremen are geographically located in close proximity to their production counterparts so that they can effectively perform the weekly schedule.		
The reliability engineering group is split. One group supports day-to-day activities whereas the other larger group works on reliability improvement.		
The planners never seem to have time to plan. They are always working to support the foremen and the crews doing the daily maintenance work.		
The maintenance manager has assigned his field superintendent to look after the day-to-day and spends the majority of his own time focused on strategic reliability work initiatives.		
The first thing that the maintenance manager does in the morning is have a meeting with production to find out what problems arose overnight. They then mobilize the organization to address these hot issues.		

Figure 11-3 Reliability is NOT a Department

Reliability is not a department.
It is a way of thinking about the plant equipment–a mind set.
As a result, production, maintenance,
and reliability engineering have equally important
and overlapping roles and responsibilities,
all to ensure that a high level of plant reliability
is maintained.

Figure 11-4 Process vs. Functional Centric Organizations

Functional centric (centered) organizations are designed with the individual job functions as their focus. These tend to be more strategic. Process centric organizations are designed with a focus on the work process. These tend to be more tactical. The Hybrid organization splits their focus and attempts to address both types.

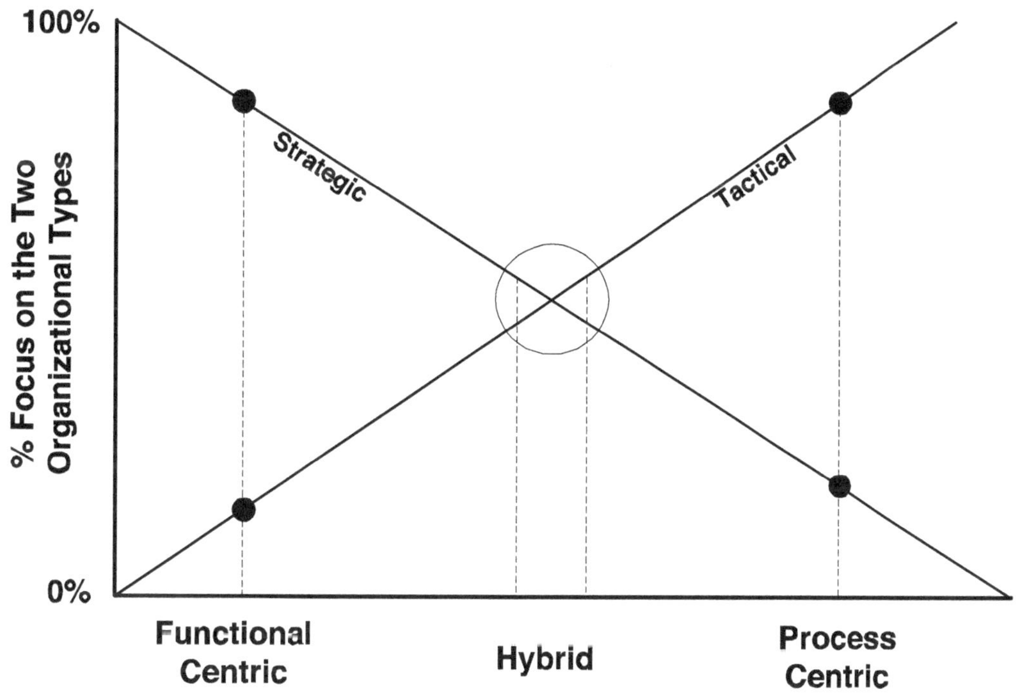

Figure 11-5 Process vs. Functional Centric Organizations

	Functional Centric Structure	Process Centric Structure
Focus	Individual organizational functions	Manufacturing process that the functions support
Positive	Ability to work outside of the day-to-day efforts on more strategic projects. Close proximity to others in the function to promote interaction and mentoring of new / less experienced personnel.	Team approach focused on getting and keeping demand of production satisfied. Close daily interaction. The team can quickly and collectively address issues.
Negative	More attention to function and peers vs. process. Process could suffer. Interrelationship problems between functions as group members find it difficult to allocate their time.	More focused on day-to-day production. The tactical is addressed at the expense of the strategic issues. Lack or reduced mentoring of functional skills. Lack of consistent approach due to a weaker functional connection.

Exercise 11-2 Structural Pitfalls

Listed below are several structural pitfalls. There are still other pitfalls not listed with which you may have had personal experience. Select from the list or use your own for this exercise. First list the results of falling into this pitfall. Next, suggest alternatives to accomplish your goal without necessarily restructuring the organization.

1. We want to make improvements in the process, so why not restructure?
2. A change in management resulting in a structural change
3. Adjustments to a structure to accommodate individuals and their specific skills
4. Recommendations from a process consultant
5. Others?

Pitfall #	**Results**
Suggested Alternative	
Suggested Alternative	
Suggested Alternative	

Figure 11-6 Structure Overlaps

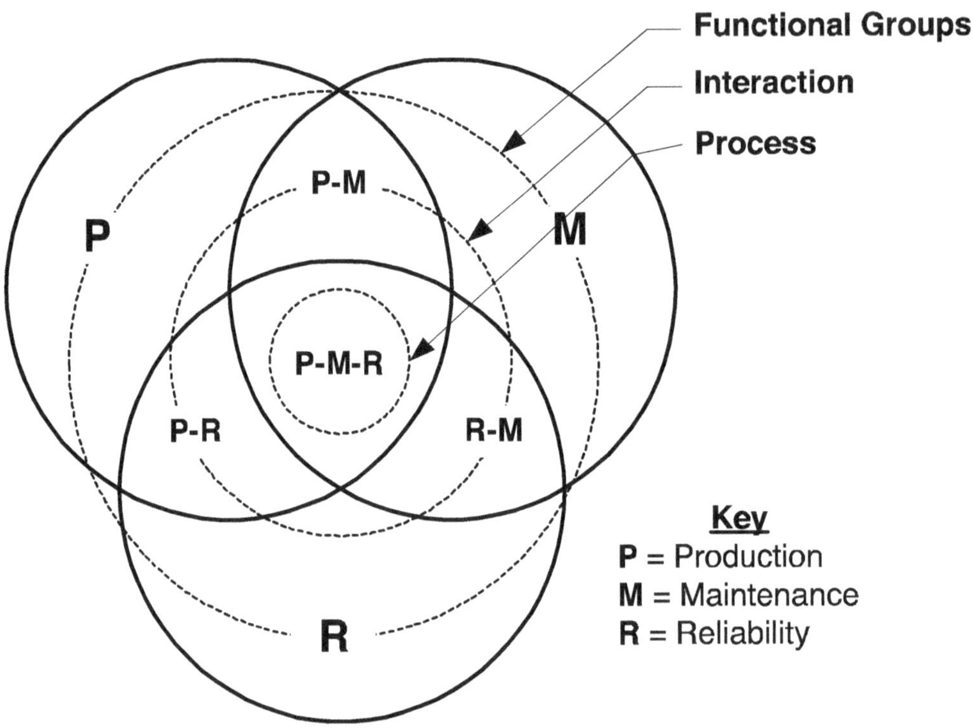

Exercise 11-3 ATPCo's Structure

Structural problems are created as a result of cultural issues. If you consider Section E.1 parts a through d and Figures A-2 through A-5 of the appendix, you will see a detailed description of ATPCo's eastern plant organization. The detailed description points to the fact that ATPCo has been set up as a hybrid organization in an attempt to bring both a process- and functional-centric strategy to the work.

However, Mike Kane has subverted the original intent of the structure with the forceful implementation of his "run it hard, fix it fast" mentality. Everyone is fixated on the day-to-day repair. The engineers, foremen, mechanics, and others throughout the organization think highly of those who are able to fix the equipment fast, they have no time for those who advocate taking time to make reliability-focused repairs. Even the work process has been altered. The scheduler handles materialization of the daily work simply because there is no schedule to prepare. Today's work isn't actually known until the day crew arrives and finds out what broke down on the previous night shift. Rewards, which include promotions, are lavished on the rapid responders; those advocating alternate courses of action are often reprimanded. In fact, Kane has a lavish network of informants so that he can keep on top of what is really happening in the plant.

With Kane's departure and a new directive from management to operate the facility with a focus on reliability, what do you need to do with the structure from a cultural perspective?

Exercise 11-3 continued

Can the existing structure be utilized to support reliability-focused values? Explain your reasoning.
Where in the structure would you expect to see the role models who could support and promote a reliability focus? Why?
Could the organization's structure support a set of rites and rituals different from those promoted by Kane? Explain your reasoning.
If you could, what would you change relative to the organizational structure, as shown in the appendix?

Exercise 11-4 Your Structure

Having diagnosed ATPCo's structure from a cultural standpoint, apply the same analysis to your own organization. Work at the reliability / maintenance level (or an even lower level) to conduct the analysis. In this manner, you will be able to apply the four elements of culture without addressing too broad of an organizational layer.

1. What is your plant's vision of maintenance and reliability?

2. Draw your organization's structure.

3. Does your structure and how it is employed support the organizational values? If not, why not? What needs to be changed?

4. Do the role models in your organization promote and support the organizational values? Are they positioned to have a positive effect? If not, why not? What needs to be changed?

5. Are the rites and rituals supported by the structure? If not, why not?

6. What elements of the cultural infrastructure exist in the organization? Where do they sit within the structure? Do they provide positive support or do they undermine your change efforts?

Exercise 11-5 Structural Web of Change

The structure section of the web of change survey is divided into four parts, as shown below. By answering the survey's 16 questions and entering your scores on the four-axis web, you will see how structure scores within the four elements of culture.

Scores:

Category	Questions	Score
Organizational Values	1 to 4	
Role Models	5 to 8	
Rites and Rituals	9 to 12	
Cultural Infrastructure	13 to 16	
Total Score – Work Process	———	

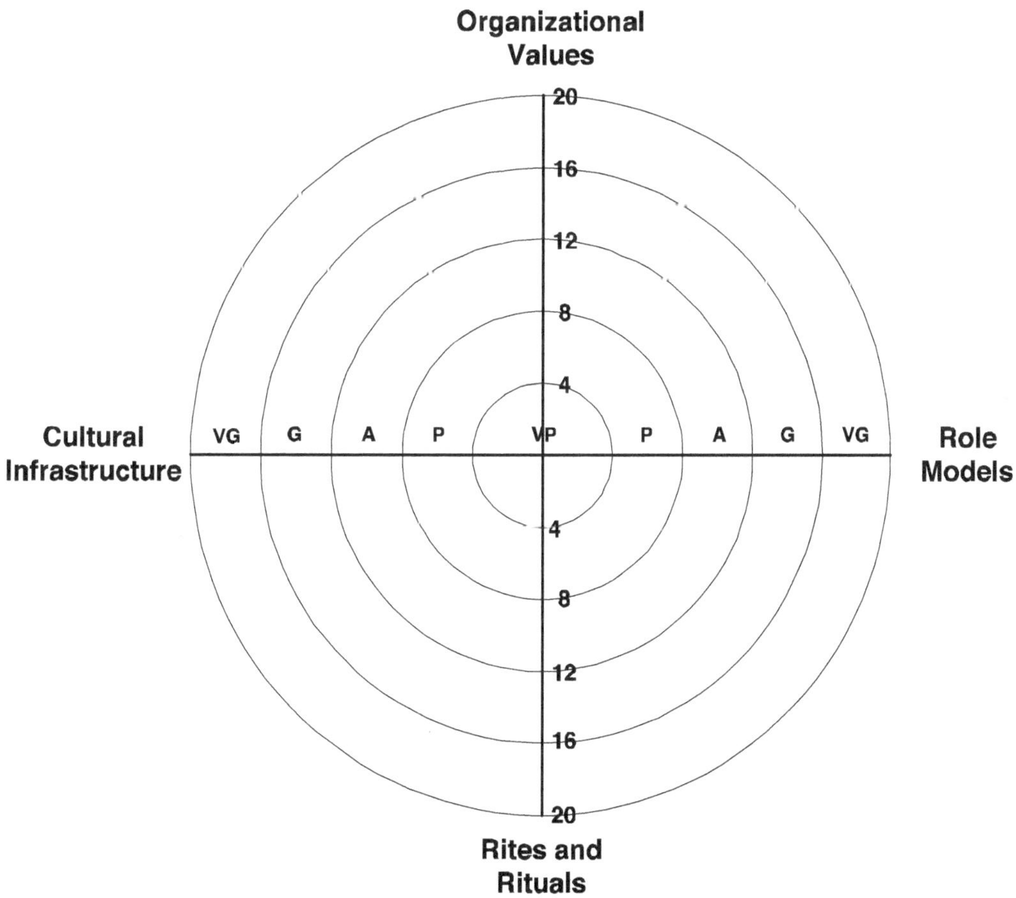

C. Structure

		Points
1	The structure has been designed to reflect the organizational values.	
	A. Strongly agree	5
	B. Agree	4
	C. Neutral	3
	D. Disagree	2
	E. Strongly disagree	1
2	Reliability and maintenance are understood to be a responsibility of multiple organizations and not a business function. This shared responsibility is clearly reflected in the values.	
	A. Strongly agree	v5
	B. Agree	4
	C. Neutral	3
	D. Disagree	2
	E. Strongly disagree	1
3	Line and staff work issues are adequately handled by the structure. The structure was designed to make this possible.	
	A. Strongly agree	5
	B. Agree	4
	C. Neutral	3
	D. Disagree	2
	E. Strongly disagree	1
4	When the structure was redesigned for the new process, the effort involved more than simply rearranging the organizational boxes. Serious thought was put into the effort.	
	A. Strongly agree	5
	B. Agree	4
	C. Neutral	3
	D. Disagree	2
	E. Strongly disagree	1

5 The structural changes were made with involvement from the leadership team and role model. It was not a case of "here is how it is going to be done," – another mandate from management.

A. Strongly agree	5
B. Agree	4
C. Neutral	3
D. Disagree	2
E. Strongly disagree	1

6 People in leadership positions within the structure are now or are becoming the role models for the organization.

A. Strongly agree	5
B. Agree	4
C. Neutral	3
D. Disagree	2
E. Strongly disagree	1

7 The current role models within the structure promote the new way things are to be done.

A. Strongly agree	5
B. Agree	4
C. Neutral	3
D. Disagree	2
E. Strongly disagree	1

8 Structural vacancies are filled with the objective of hiring individuals who will ultimately become role models and promote the new process.

A. Strongly agree	5
B. Agree	4
C. Neutral	3
D. Disagree	2
E. Strongly disagree	1

9 The structure supports the rites and rituals as dictated by the work process.

A. Strongly agree	5
B. Agree	4
C. Neutral	3
D. Disagree	2
E. Strongly disagree	1

10 Roles and responsibilities are clear for everyone within the structure, specifically how they support the rites and rituals associated with the process.

A. Strongly agree	5
B. Agree	4
C. Neutral	3
D. Disagree	2
E. Strongly disagree	1

11 The former set of rites and rituals (those being replaced with the new process) have been removed from how things are done around here and audits are conducted to assure this fact.

A. Strongly agree	5
B. Agree	4
C. Neutral	3
D. Disagree	2
E. Strongly disagree	1

12 The work locations of those in the organizational structure support the rites and rituals of the process.

A. Strongly agree	5
B. Agree	4
C. Neutral	3
D. Disagree	2
E. Strongly disagree	1

13 The members of the cultural infrastructure whose roles have been altered by the change have been identified, and efforts have been made to smooth this transition.

A. Strongly agree	5
B. Agree	4
C. Neutral	3
D. Disagree	2
E. Strongly disagree	1

14 A structural change plan was coupled with effective and efficient communication and used to minimize the effect of the gossips.

A. Strongly agree	5
B. Agree	4
C. Neutral	3
D. Disagree	2
E. Strongly disagree	1

15 A structural change plan was coupled with effective and efficient communication and used to minimize the effect of the whisperers.

A. Strongly agree	5
B. Agree	4
C. Neutral	3
D. Disagree	2
E. Strongly disagree	1

16 The Keepers of the Faith have been included in the structural change process and enlisted to support it by their cultural infrastructure involvement.

A. Strongly agree	5
B. Agree	4
C. Neutral	3
D. Disagree	2
E. Strongly disagree	1

Exercise 11-6 The Structure Web – Analyzing the Low Spots

This exercise will help you use the structure web survey you completed in Exercise 11-5 to address potentially-needed corrective actions within the four elements of culture.

Scores:

Category	Questions	Score	VP 1-4	P 5-8	N 9-12	G 13-16	VG 17-20
Organizational Values	1 to 4						
Role Models	5 to 8						
Rites & Rituals	9 to 12						
Cultural Infrastructure	13 to 16						

Enter your scores from the work process web survey in the "Score" column. Then mark an X in the appropriate column, based on the following key. Scores are classified as: VP = very poor (1 – 4 points); P = poor (5 – 8 points); A = average (9 – 12 points); G = good (13 – 16 points); and VG = very good (17 – 20 points)

2. Questions:

Which cultural scores are low (points between 1 and 8)?
Why do you believe that these cultural areas have scored this low?
From a structure perspective, what could be done to make positive improvement?
Which cultural scores are high (points between 13 and 20)?
Why do you believe that these cultural areas have scored this high?
From a structural perspective, what could be done to generate improvements in the low scoring areas?

Group Learning

Chapter 12

Purpose

This chapter introduces the fourth of the eight elements of change—group learning. This element is important because learning is the way we evolve and improve our performance.

What Is Included

Figure 12-1 Group Learning Defined

Figure 12-2 Filters

Exercise 12-1 Filters and What You Can Miss

Figure 12-3 Type I and II Learning Models

Exercise 12-2 Identifying Type I and II Learning

Figure 12-4 "S" Curves

Exercise 12-3 "S" Curve Examples

Figure 12-5 Spiral Learning

Exercise 12-4 Your Spiral Learning Examples

Figure 12-6 The Learning vs. the Blaming Organization

Exercise 12-5 The Learning vs. the Blaming Organization

Exercise 12-6 Group Learning and the Four Elements of Culture

Exercise 12-7 The Learning Web of Change

Exercise 12-8 The Learning Web – Analyzing the Low Spots

Figure 12-1 Group Learning Defined

Group learning is the ability of individuals to acquire new knowledge, then use this knowledge in a group setting that either leads to aligned alternative courses of action or reinforces that the current course of action is correct.

Figure 12-2 Filters

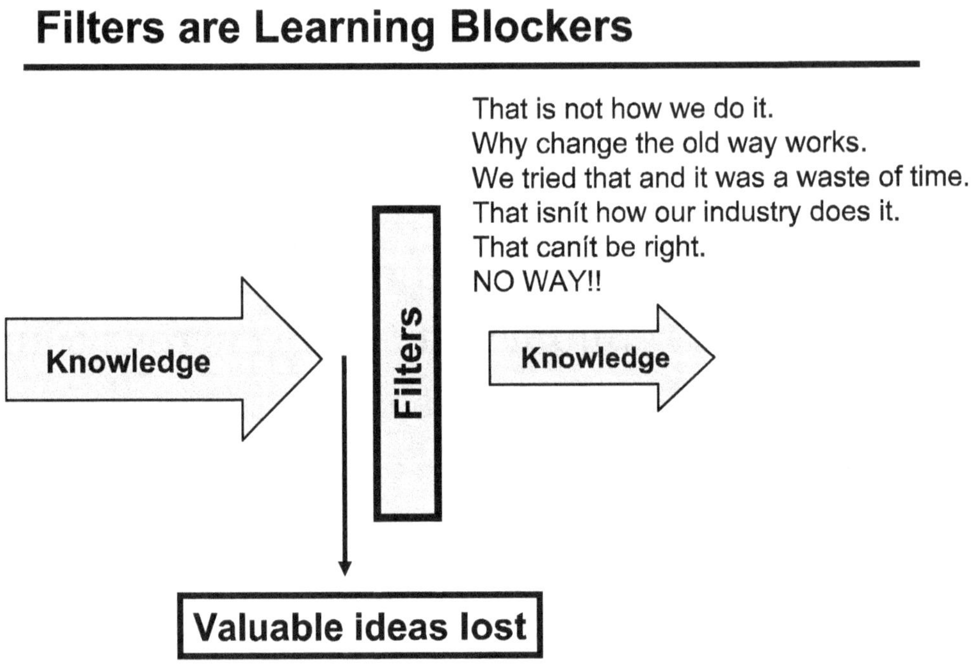

Exercise 12-1 Filters and What You Can Miss

One of the problems we face as we attempt to learn is that, individually and collectively, we filter information. We have difficulty allowing new ideas past our filters so that we can analyze the information and determine if it is applicable. If we fail to turn off the filters, we are doing a disservice to our company by potentially missing new ways to improve the work. Filters of this type look and sound like:

- That is not how we have done this work before.
- We have been successful with the old methodwhy change?
- We tried that before and it was a waste of time.
- Our way of doing _____ is better.
- That is a different industryit won't work in ours.
- That can't be right; our information says something different.

What are some of the filters you encounter? How have they hurt you and your organization's ability to learn?

Filter	How has this filter hampered learning?

Exercise 12-1 continued

For the same filters, what could be done to turn them off so that group learning can take place?

Filter	How can this filter be turned off?

Figure 12-3 Type I and II Learning Models

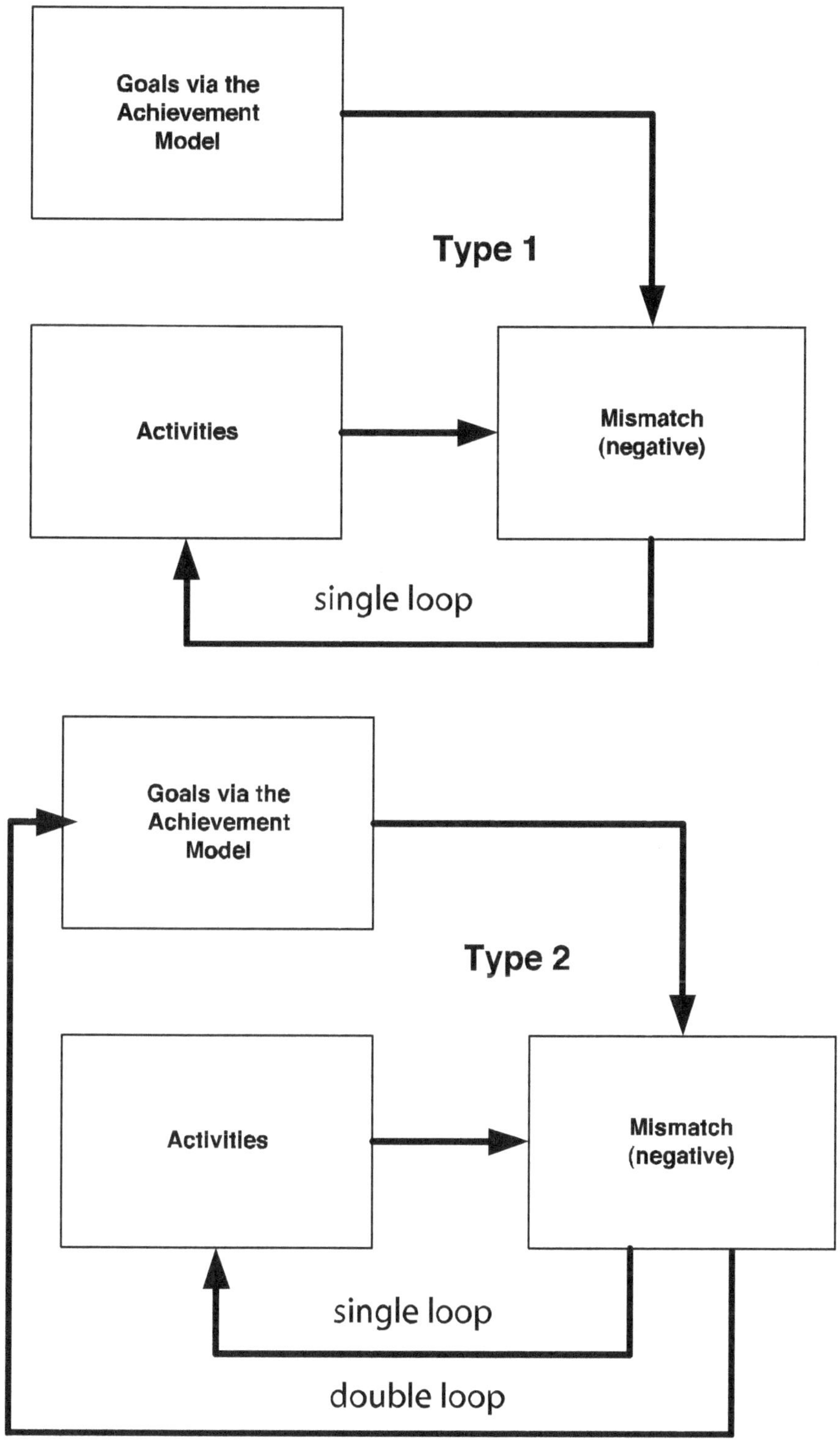

Exercise 12-2 Identifying Type I and II Learning
Part 1

Figure 12-3 shows the two types of organizational learning. In Part 1 of this exercise, review each representative scenario, then determine if it is describing Type I or Type II learning. Mark an "X" in the column that describes the type of learning taking place.

Scenario	Type I?	Type II?
The organization recognizes that its maintenance activities are focused on rapid repair, but should be focused on failure prevention.		
The measurement that shows the level of compliance to the preventive maintenance program indicates that the program is behind schedule. As a result, more mechanics are assigned to get caught up.		
The number of unplanned work orders in the backlog has increased beyond what management believed was a reasonable number. Additional planners were assigned to get the backlog of unplanned jobs under control.		
The maintenance work orders being planned were reviewed and found generally to be deficient. As a result, a training program for planners was instituted to improve the quality of their work.		
Production had recently complained about the reliability engineers not providing reliability improvement recommendations for several of the pump repair jobs in the shop. When management looked into the allegation, they discovered that the engineers were too busy working on the day-to-day work in the field. Their focus was immediately changed.		
Safety work orders were not getting completed in the 30-day window promised by the maintenance organization. The workforce was scheduled to complete many of them on overtime.		
Safety work orders were not getting completed in the 30-day window promised by the maintenance organization. The production and maintenance organizations reviewed the work orders. They identified and addressed the root causes so that these problems would not reoccur in the future.		

Part 2

As a second part of this exercise, write the scenarios from Part 1 that are Type I. What could be done with these to perform a type II analysis?

Scenario that fits Type I	What could be done with this to conduct a Type II analysis?

Figure 12-4 "S" Curves

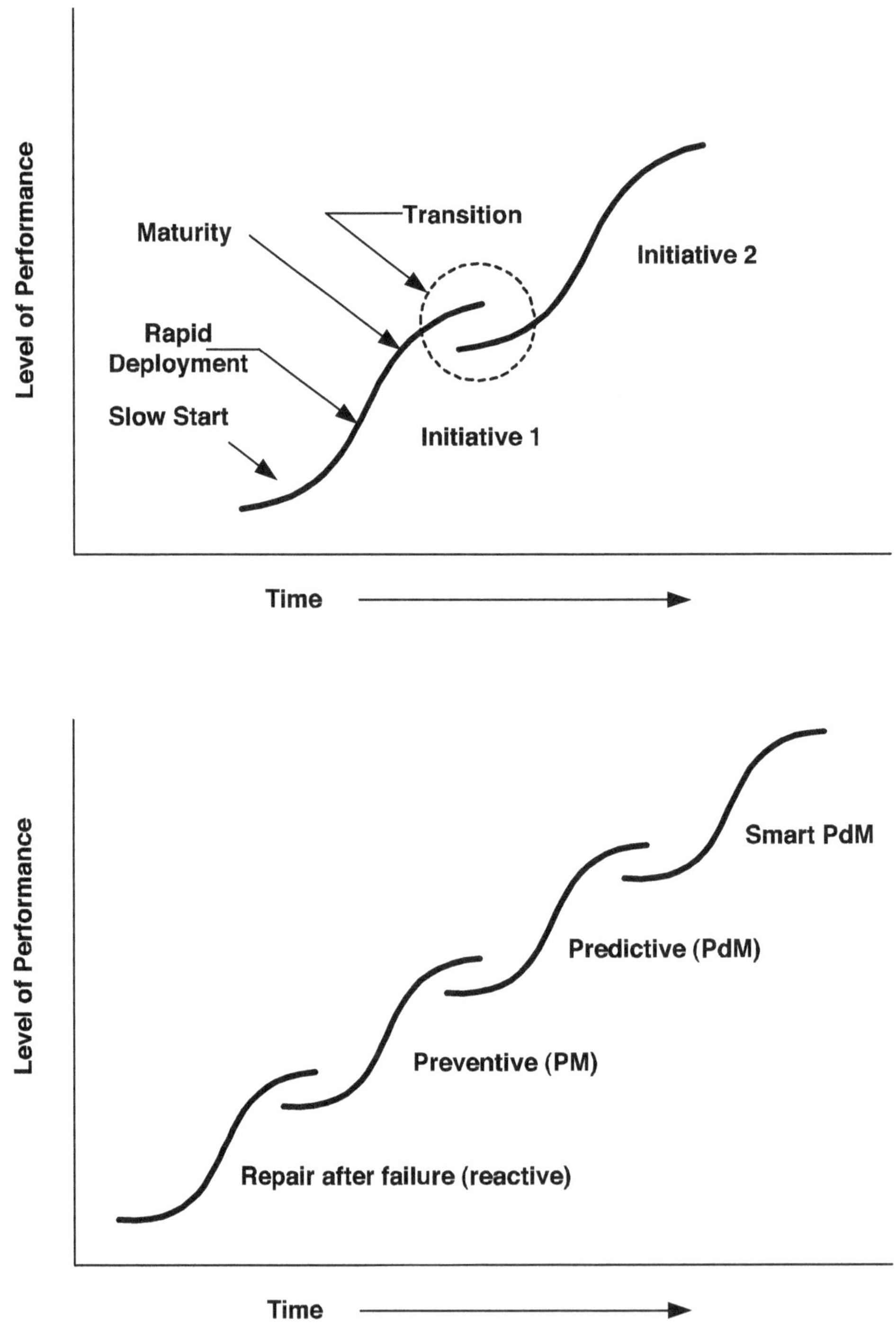

Exercise 12-3 "S" Curve Examples
Part 1

Draw an "S" curve of your own, describing a process that fits the "S" curve model.

Part 2

Describe two stages of your "S" curve process.

What was the initiative?
Describe the start of the effort.
What did maturity look and feel like?
What was the second initiative?
How was the transition accomplished?

Figure 12-5 Spiral Learning

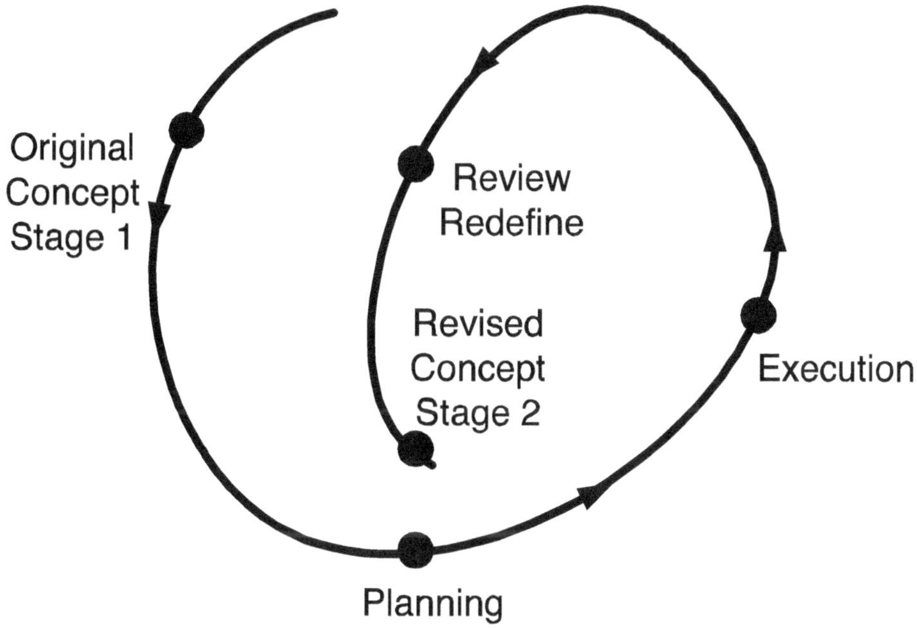

Exercise 12-4 Your Spiral Learning Example

Based on your knowledge of spiral learning, describe a spiral learning process of your own.

What was the initial concept?
What planning did you do in order to put the concept into place?
What did execution look like?
What did you learn in the review and redefine phase?
What did the second spiral look like?
Did the second spiral have elements included that you did not know until you completed the first loop? If so, what were they?

Figure 12-6 The Learning vs. the Blaming Organization Part 1

Learning and Blame

Two types of organizations

The learning organization → **Search for the root cause and correct it.**

The blaming organization → **Search for the guilty and punish them.**

Figure 12-6 The Learning vs. the Blaming Organization Part 2

Types of Actions

Inadvertent Actions

Inadvertent actions are steps that we did because we believed that they were correct. They are usually the last link in a long chain of failures that led up to the event.

Willful Actions

Willful actions are steps that we did even though we knew at the time that what we were doing was wrong.

Exercise 12-5 The Learning vs. the Blaming Organization

Blaming organizations lose valuable information and inhibit their ability to identify and implement corrective action based on true root causes. On the other hand, learning organizations approach a failure in a different manner, permitting people to provide valuable root cause insight without fear of retribution.

Review the following scenarios and identify how you and your organization would react. Do you fit the learning or the blaming organizational profile?

1. A critical pump was recently overhauled by maintenance. On start up, itimmediately failed, resulting in a full day of lost production while it was repaired a second time.

How would you and your organization react?

Are you a learning or blaming organization in this case?

What is gained or lost because of your approach?

If you recognize that there is value altering your approach in this case, what would you do differently and why?

2. You receive an emergency phone call at 10:00 AM informing you that the main tractium production line has just shut down. When you arrive at the line the situation is serious and there is no one around giving directions to correct it. Your lead operator is nowhere to be found.

How would you and your organization react?

Are you a learning or blaming organization in this case?

What is gained or lost because of your approach?

If you recognize that there is value altering your approach in this case, what would you do differently and why?

3. One of your mechanics is injured when a tool left behind from a prior shift fell from a scaffold and struck him on the shoulder.

How would you and your organization react?

Are you a learning or blaming organization in this case?

What is gained or lost because of your approach?

If you recognize that there is value altering your approach in this case, what would you do differently and why?

4. A structural pipe support fails, causing several live lines to become dislodged. One of them opens at a pair of flanges, causing a serious spill of product to the ground. The cost, time to repair, and environmental impact are significant. You discover that one of your new engineers was the designer.

How would you and your organization react?

Are you a learning or blaming organization in this case?

What is gained or lost because of your approach?

If you recognize that there is value altering your approach in this case, what would you do differently and why?

5. Two mechanics are caught leaving the plant early without permission. When you do some preliminary investigation, you also find out that the job they were doing was left incomplete. They had told their supervisor it was finished.

How would you and your organization react?

Are you a learning or blaming organization in this case?

What is gained or lost because of your approach?

If you recognize that there is value altering your approach in this case, what would you do differently and why?

Exercise 12-6 Group Learning and the Four Elements of Culture

1. What is your plant's vision for its reliability and maintenance functions?

2. Organizational Values

a) How have the organizational values supported the organization in learning and achieving the vision?

b) What was done initially to help the organization learn of the values?

c) What has been done to sustain this over time as well as apply Type II learning?

d) What do you see as your opportunities for improvement regarding helping the organization learn and apply the organizational values?

3. Role Models

a) How have the existing role models helped the organization learn and apply the values?

b) If they have not supported this learning, why not?

c) How can the existing or new role models help the organization learn and apply the organizational values?

4. Rites and Rituals

a) What are the existing rites and rituals associated with how your organization addresses reliability and maintenance?

b) How were they learned?

c) If you plan to change your organization's focus from reactive to proactive maintenance, how would you change your rites and rituals?

5. Cultural Infrastructure: Describe how each of the roles within the cultural infrastructure contribute to group learning within your organization.

a) Story Tellers

b) Keepers of the Faith

c) Whisperers

d) Gossips

e) Spies

f) Symbols

g) Language

Exercise 12-7 The Group Learning Web of Change

The group learning section of the web of change survey is divided into four parts, as shown below. By answering the survey's 16 questions and entering your scores on the four-axis web, you will see how group learning scores within the four elements of culture.

Scores:

Category	Questions	Score
Organizational Values	1 to 4	
Role Models	5 to 8	
Rites and Rituals	9 to 12	
Cultural Infrastructure	13 to 16	
Total Score – Group Learning	———	

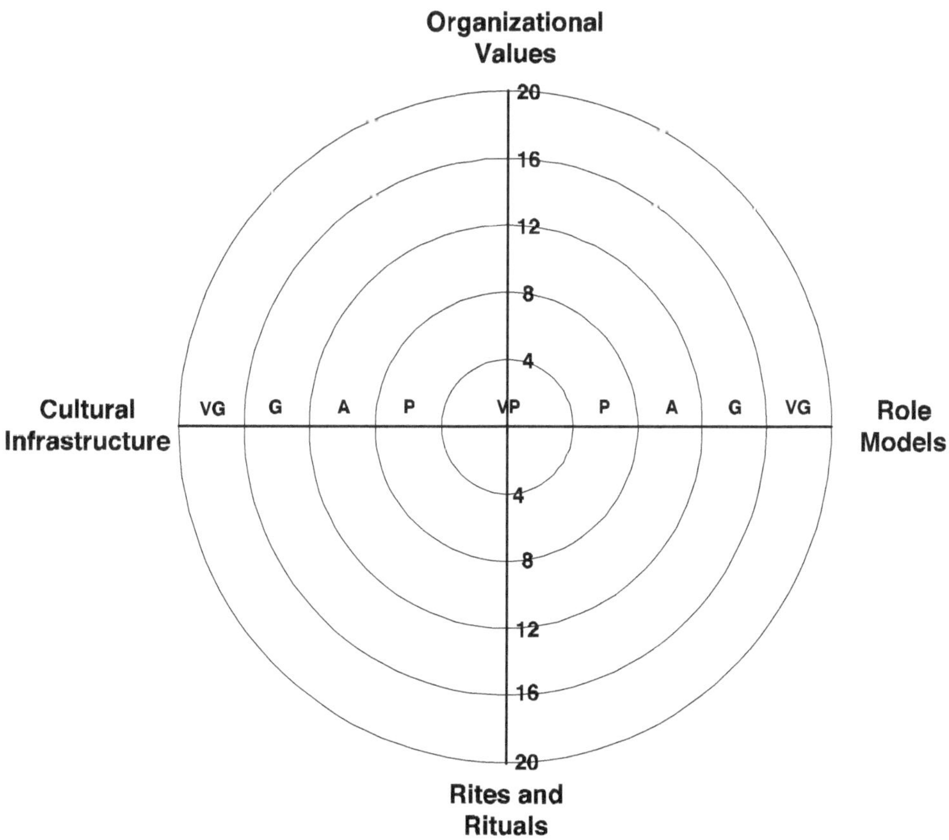

D. Group Learning

1 Type 1 (single loop) learning is used to support current goals.

	Points
A. Strongly agree	5
B. Agree	4
C. Neutral	3
D. Disagree	2
E. Strongly disagree	1

2 The company or site is open to Type 2 (double loop) learning even though employing it may identify that its goals are incorrect.

	Points
A. Strongly agree	5
B. Agree	4
C. Neutral	3
D. Disagree	2
E. Strongly disagree	1

3 Gaps identified in either of the two types of learning processes are handled by developing corrective action items and following up to completion.

	Points
A. Strongly agree	5
B. Agree	4
C. Neutral	3
D. Disagree	2
E. Strongly disagree	1

4 One of the organizational values is the need to learn. The process is in place to assure that learning is continuously taking place.

	Points
A. Strongly agree	5
B. Agree	4
C. Neutral	3
D. Disagree	2
E. Strongly disagree	1

5 Role models are in place who understand the organization's need to learn.

A. Strongly agree	5
B. Agree	4
C. Neutral	3
D. Disagree	2
E. Strongly disagree	1

6 The role models actively support the learning process and, where it is not working, they encourage it.

A. Strongly agree	5
B. Agree	4
C. Neutral	3
D. Disagree	2
E. Strongly disagree	1

7 Role models recognize the need to use the Goal Achievement Model (or similar tool) to promote learning while at the same time accomplishing the vision, goals, and initiatives.

A. Strongly agree	5
B. Agree	4
C. Neutral	3
D. Disagree	2
E. Strongly disagree	1

8 When new things are learned that could have a positive affect on the business of reliability and maintenance, role models are open to trying them out instead of throwing up barriers that block new ideas.

A. Strongly agree	5
B. Agree	4
C. Neutral	3
D. Disagree	2
E. Strongly disagree	1

9 Rites and rituals are in place and support the learning spiral.

A. Strongly agree	5
B. Agree	4
C. Neutral	3
D. Disagree	2
E. Strongly disagree	1

10 Rites and rituals are focused on supporting a learning organization, not a blaming one.

A. Strongly agree	5
B. Agree	4
C. Neutral	3
D. Disagree	2
E. Strongly disagree	1

11 Rites and rituals have embedded within them processes to identify gaps, learn from those gaps, and take the necessary corrective action.

A. Strongly agree	5
B. Agree	4
C. Neutral	3
D. Disagree	2
E. Strongly disagree	1

12 Rites are used as a learning and reinforcing process for the rituals. When necessary, rituals are modified based on what was learned.

A. Strongly agree	5
B. Agree	4
C. Neutral	3
D. Disagree	2
E. Strongly disagree	1

13 The cultural infrastructure is monitored to remove learning barriers.

A. Strongly agree	5
B. Agree	4
C. Neutral	3
D. Disagree	2
E. Strongly disagree	1

14 Active efforts are in place to promote group learning within the organization, especially within the cultural infrastructure.

A. Strongly agree	5
B. Agree	4
C. Neutral	3
D. Disagree	2
E. Strongly disagree	1

15 Language and symbols are being learned that represent the new process.

A. Strongly agree	5
B. Agree	4
C. Neutral	3
D. Disagree	2
E. Strongly disagree	1

16 The Keepers of the Faith are being enlisted to mentor the learning process within the site.

A. Strongly agree	5
B. Agree	4
C. Neutral	3
D. Disagree	2
E. Strongly disagree	1

Exercise 12-8 The Group Learning Web – Analyzing the Low Spots

This exercise will help you use the group learning web survey you completed in Exercise 12-7 to address potentially-needed corrective actions within the four elements of culture.

Enter Your Group Learning Survey Scores.

Scores:

Category	Questions	Score	VP 1-4	P 5-8	N 9-12	G 13-16	VG 17-20
Organizational Values	1 to 4						
Role Models	5 to 8						
Rites & Rituals	9 to 12						
Cultural Infrastructure	13 to 16						

Enter your scores from the work process web survey in the “Score” column. Then mark an X in the appropriate column, based on the following key. Scores are classified as: VP = very poor (1 – 4 points); P = poor (5 – 8 points); A = average (9 – 12 points); G = good (13 – 16 points); and VG = very good (17 – 20 points)

2. Questions:

Which cultural scores are low (points between 1 and 8)?
Why do you believe that these cultural areas have scored this low?
From a group learning perspective, what could be done to make positive improvement?
Which cultural scores are high (points between 13 and 20)?
Why do you believe that these cultural areas have scored this high?
From a group learning perspective, what could be done to generate improvements in the low scoring areas?

Chapter 13

Technology

Purpose

This chapter introduces the fifth element of the eight elements of change—technology. In our context, technology refers to the software applications utilized by the reliability and maintenance organization. These applications enable change by providing the tools that are needed in areas such as; planning and scheduling, asset management, and detailed reliability focused analysis.

What Is Included

Figure 13-1	Technology's Involvement with Change
Figure 13-2	The Key Factors of Your Technology
Exercise 13-1	Analyzing Your Technology (High-Level View)
Figure 13-3	Rules for Information (Book Reference Page 197)
Exercise 13-2	Technology in Support of the Other Elements of Change
Exercise 13-3	Technology and the Four Elements of Culture
Exercise 13-4	The Technology Web of Change
Exercise 13-5	The Technology Web – Analyzing the Low Spots

Figure 13-1 Technology's Involvement with Change

Working effectively and efficiently in a reliability-focused work environment requires timely and accurate information. Meeting this requirement is where technology fits into the picture, both in the systems we employ and the information that they deliver. As used in our discussion, technology is the enabler of change. It supports the other seven elements of change as well as providing a reinforcing foundation for the four elements of culture.

Figure 13-2 The Key Factors of Your Technology

Accurate

The information within our systems needs to be accurate. Older systems or manual efforts do not always live up to this requirement. Older systems are difficult to manage and information doesn't always get entered. Both manual systems and older computer systems that require extensive manual intervention are subject to human error.

Timely

Reliability-based decisions require information to be at people's finger tips when needed.

Controlled

The information within our systems needs to be available, but it also needs to be controlled. The majority of people who need information really only need to see it; they should not be able to remove, alter, or delete it.

Widely Accessible

If we want the work to have a reliability focus, everyone involved must have access to the information that will lead them to quality decisions.

Consistently Logical

Information has a logic to it. Getting to the correct information follows a logical sequence and is user friendly. Additionally, once the information is obtained about an asset, it makes sense. Software should provide consistently logical results.

Integrated

If you want people to make quality decisions, then you need systems that allow them to gain access to all of the information required to make those decisions. You need to establish a way to bring all of the information together for access by those who require it—one that links the systems and the information that these systems contain.

Exercise 13-1 Analyzing Your Technology (High-Level View) Part 1

Figure 13-2 presented six key factors that relate technology to the work of reliability and maintenance. In this exercise, you will evaluate your reliability / maintenance technology based on these six factors. When you are evaluating your technology, consider all of the applications that are employed in the service of improved reliability and maintenance in your plant.

Your Evaluation: (See Part 2 of this exercise before you score yourself.)

Category	Score	Notes
Accurate		
Timely		
Controlled		
Widely Accessible		
Consistently Logical		
Integrated		

Scores are: very poor = 1, poor = 2, average = 3, good = 4 and very good = 5

Part 2 Things To Think About as You Analyze the Element of Technology

These questions apply to information databases, manual systems, or combinations thereof.

Accurate

1. How many software applications contain the same asset-related information. Is the information consistent?
2. Who inputs new information or information that changes? Is it an assigned task for an individual or group or is it everyone's job? How well is it done?
3. Is there a detailed procedure for adding or changing information? If so, is it followed?
4. Is the information in your databases ever checked for accuracy? If it is not correct, is there a process to get it corrected?

Exercise 13-1 (continued)

Timely

1. Are additions and / or changes to the information made in a timely fashion.
2. Do those who need information know how to get it without long delays?

Controlled

1. Is there a process in place where only select (trained) individuals can add, change, or delete information in the asset database?
2. Is there a procedure for adding, changing, or deleting information so that the correct information is placed into the various systems?
3. Are the information databases backed up so that the information can not be lost?

Widely Accessible

1. Is the information about your plant assets widely accessible to those who need it? This could possibly be provided by "read only" access to the various applications.
2. Do the users of the information have to have access to multiple applications and know how to us them to get information?

Consistently Logical

1. Is there a logical hierarchy or file structure for your asset-related information?
2. Is there a logical structure within each asset file so that information can be found?
3. Are they followed by those who add, change, or access the information?
4. When people remove information (especially in paper systems), is there a check-in / check-out process so that you know who has the information and that they return it?

Integrated

1. Are the information databases integrated or interfaced so that a change in one is reflected in all of the others?
2. Is your system set up so that you only need to make an addition or change in one place and it is automatically captured by the other systems?

Exercise 13-1 (continued)

Part 3 Chart Your Technology Evaluation

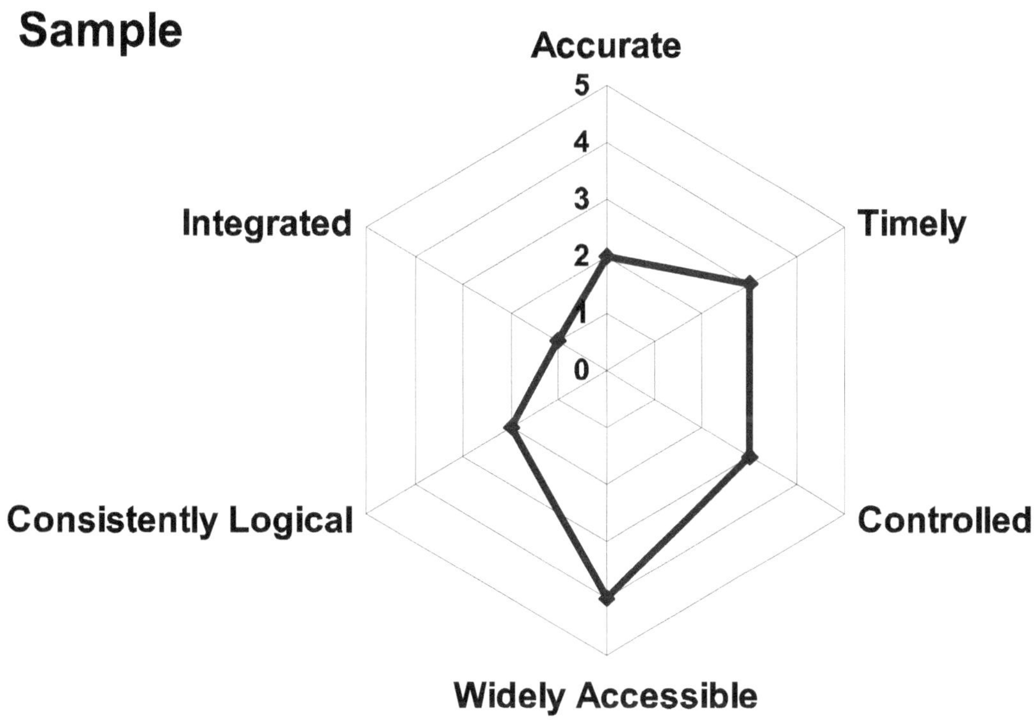

Your Web

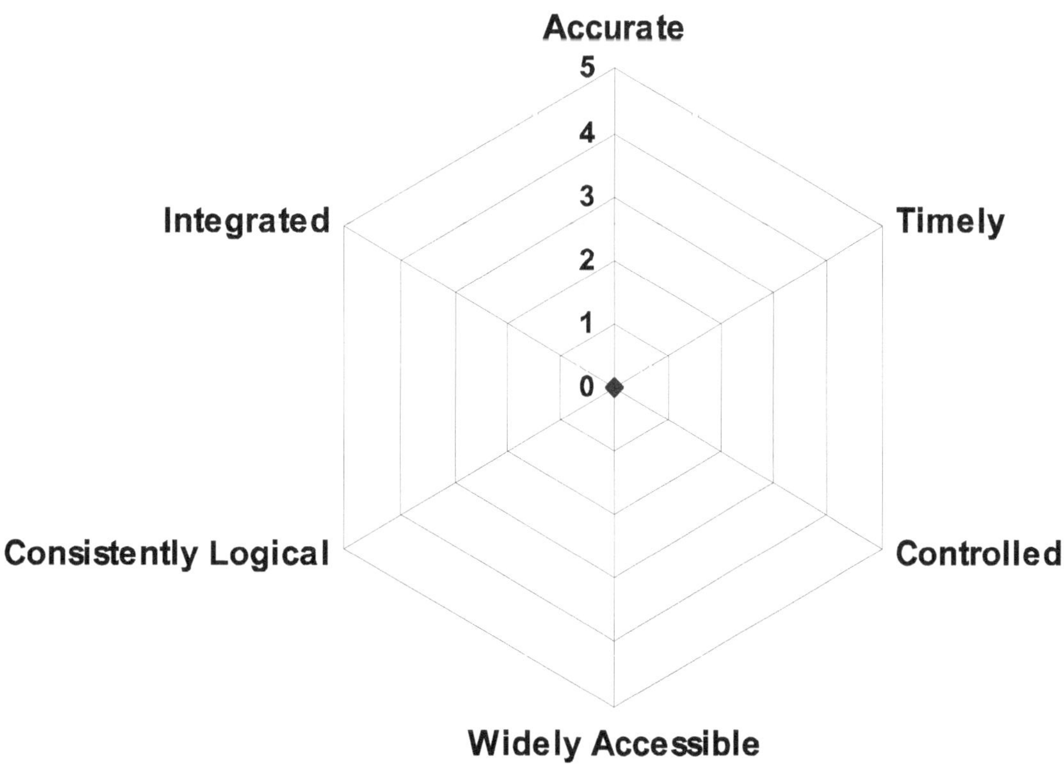

Exercise 13-1 (continued)

Part 4 Chart Analysis

Which elements scored poor or very poor?
For each element that scored poor or very poor, list the reasons you believe that they received this score.
What do you believe can be done to correct the problems identified?

Figure 13-3 Rules for Information (Book Reference Page 197)

1. Never have the same information in more than one system.

2. Make information accessible in a read-only mode and an easy-to-obtain format. (Integrate, if possible.)

3. Tightly control who can add, change, or delete information.

4. Have a detailed process in place for new information, information changes, and information removal.

5. Make certain that the information handling process is followed.

6. Evaluate new approaches to information handling and access.

Exercise 13-2 Technology in Support of the Other Elements of Change

Technology plays an important role in the change process and, therefore, it is important to recognize how technology impacts the other seven elements of change.

Part 1

How the Elements Fit with Technology

Element	How it Fits
Leadership	Implementing a new technology or optimally using the ones you have requires a vision, goals, initiatives, and specific activities (the Goal Achievement Model) in order to mobilize the organization. Leadership sets this direction.
Work Process	Once an organization determines that they want to have a high-quality information system, they need to alter the work processes so that they support and are supported by the technology.
Structure	With high-quality technology, people can now interact in different ways. In the past, people needed to be housed in close proximity to one another. Now they can work from remote locations and still be closely connected.
Group Learning	As the technology improves, more and more information is readily available for accurate reliability-based decisions. By making decisions with all of the information, groups and individuals learn to do things better.
Communications	Technology promotes communication of information between groups and individuals in a timely manner.
Interrelationships	Technology improves interrelationships between groups and people because needed information is easy to obtain and share.
Rewards	There is nothing more frustrating than needing information and not being able to obtain it. Technology promotes rewards at many levels within the Hierarchy of Needs.

Part 2

Scenarios for this Exercise: This exercise lists seven scenarios. Identify which elements of change are affected by circling them; then explain why.

Leadership	Work Process	Structure	Group Learning	Communication	Interrelationships	Rewards

Scenario With the new technology recently implemented, providing wide access to the inspection database, management decided that the Inspection Group should be moved so that each inspector resided in the same office area as the other members of their work teams. This was far different than the original configuration where all of the inspectors and their files resided in the same building, far away from the work teams.

Why does it address the elements you marked?

Leadership	Work Process	Structure	Group Learning	Communication	Interrelationships	Rewards

Scenario In the past if you needed information from the machinery asset database, one of the machinery engineers had to obtain it for you. Because the engineers were very busy and most often out in the plant, getting timely information was very difficult. As a result, there was often open conflict between the engineers and those needing information.

Why does it address the elements you marked?

Leadership	Work Process	Structure	Group Learning	Communication	Interrelationships	Rewards

Scenario - A consultant hired to conduct an assessment of the plant's asset-related information identified many areas for improvement. The main finding was that there were too many databases and file systems that were not well maintained. The consultant's recommendation for an initiative to correct this problem was dismissed as too expensive.

Why does it address the elements you marked?

Leadership	Work Process	Structure	Group Learning	Communication	Interrelationships	Rewards

Scenario – Following the completion of a work initiative to improve the asset-related plant technology, management noticed that the number of failures was decreasing. When they inquired as to what was causing this, they were told that now people could access past history and use that information to make better reliability-focused decisions about upcoming work.

Why does it address the elements you marked?

Leadership	Work Process	Structure	Group Learning	Communication	Interrelationships	Rewards

Scenario – Bob, the foreman of the machine shop, recognized that Frank, one of his junior engineers, had just saved the plant considerable downtime and repair expense. It seems that Frank had checked the machinery database and discovered that some of the clearances being used for the repair of a critical pump were wrong. He found this information in a report prepared by his predecessor from the last repair. Bob was so pleased he called the plant manager.

Why does it address the elements you marked?

Leadership	Work Process	Structure	Group Learning	Communication	Interrelationships	Rewards

Scenario – Recently, one of the more senior work planners was overheard telling one of the more junior planners how to use maintenance management system tools to find information about a plan he had prepared for a similar job. At the end of the discussion, the senior planner said, "Years ago we could never have found that information. We would have had to create the plan as if it were a new event."

Why does it address the elements you marked?

Leadership	Work Process	Structure	Group Learning	Communication	Interrelationships	Rewards

Scenario – The new planning effort incorporated input from the reliability group so that equipment was repaired with the idea that future failure was to be eliminated. The reliability database provided a good deal of the information needed, including recommendations for repairs from prior jobs.

Why does it address the elements you marked?

Exercise 13-3 Technology and the Four Elements of Culture

1. **What is your plant's vision for its reliability and maintenance func tions? How has technology been used to support it?**

2. **Organizational Values**

   a) How have the organizational values been supported by your technology in order to achieve the vision?

   b) Where are the gaps? What do you see as your opportunities for improvement?

   c). What has been done to sustain / improve the use of technology over time?

3. **Role Models**

a) How have the existing role models helped the organization learn and apply the technology?

b) If they have not supported this learning, why not?

c) How can the new or existing role models help drive the use of technology in reliability-based decision making?

4. **Rites and Rituals**

a) What are the existing rites and rituals associated with how your organization utilizes technology to address reliability and maintenance?

b) If you plan to change your organization's focus from reactive to proactive maintenance, how would you change your rites and rituals associated with technology?

5. Cultural Infrastructure: Describe how each of the roles within the cultural infrastructure contribute or fail to contribute to the use of technology with in your organization.

a) Story Tellers

b) Keepers of the Faith

c) Whisperers

d) Gossips

e) Spies

f) Symbols

g) Language

Exercise 13-4 The Technology Web of Change

The technology section of the web of change survey is divided into four parts, as shown below. By answering the survey's 16 questions and entering your scores on the four-axis web, you will see how technology scores within the four elements of culture.

Scores:

Category	Questions	Score
Organizational Values	1 to 4	
Role Models	5 to 8	
Rites and Rituals	9 to 12	
Cultural Infrastructure	13 to 16	
Total Score – Technology	———	

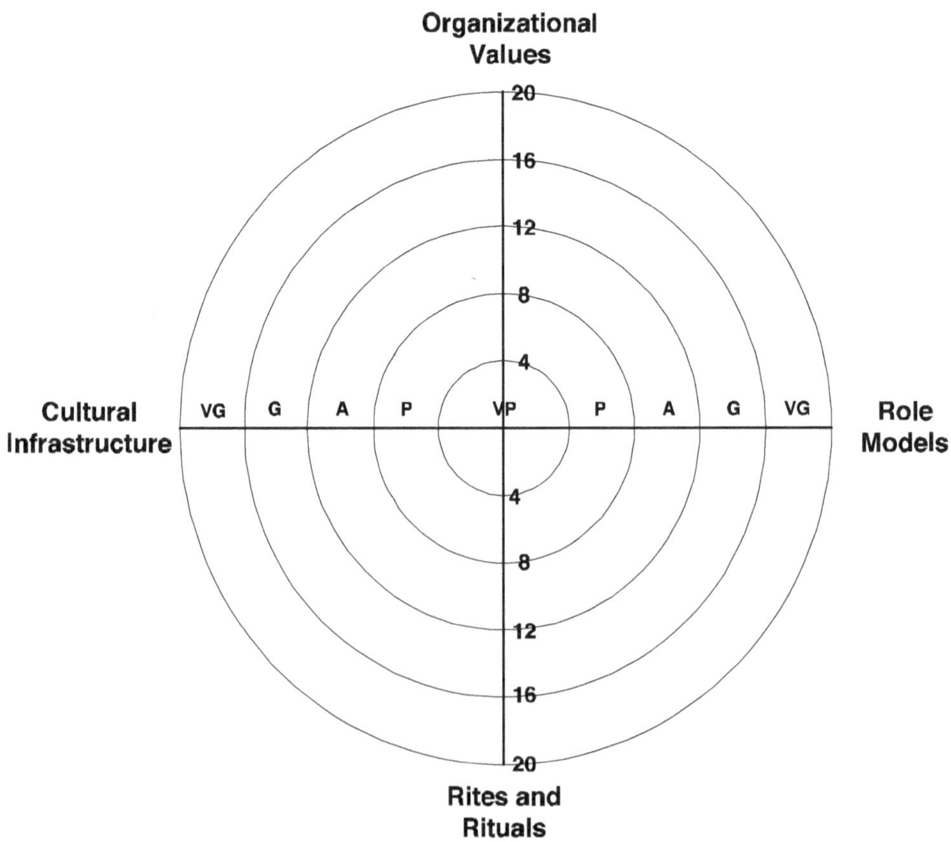

E. Technology **Points**

1 The technology has functionality that supports the organizational values.

A. Strongly agree	5
B. Agree	4
C. Neutral	3
D. Disagree	2
E. Strongly disagree	1

2 When decisions need to be made, the technology readily provides the needed data to the users.

A. Strongly agree	5
B. Agree	4
C. Neutral	3
D. Disagree	2
E. Strongly disagree	1

3 There is a close tie between the continuous improvement plan for the organization and the technology needed to support it.

A. Strongly agree	5
B. Agree	4
C. Neutral	3
D. Disagree	2
E. Strongly disagree	1

4 When the organizational values are upgraded (or changed), the technology, a critical component, is also upgraded.

A. Strongly agree	5
B. Agree	4
C. Neutral	3
D. Disagree	2
E. Strongly disagree	1

5 When role models require information to support the work, the technology makes it available

A. Strongly agree	5
B. Agree	4
C. Neutral	3
D. Disagree	2
E. Strongly disagree	1

6 The technology provides reports related to the health of the work process; they are easily obtained and can be customized to the needs of the users.

A. Strongly agree	5
B. Agree	4
C. Neutral	3
D. Disagree	2
E. Strongly disagree	1

7 Role models actively support technology solutions in support of business needs.

A. Strongly agree	5
B. Agree	4
C. Neutral	3
D. Disagree	2
E. Strongly disagree	1

8 Role models promote data integration (a single pathway to all required information available to those who need it) instead of functional / technological information silos.

A. Strongly agree	5
B. Agree	4
C. Neutral	3
D. Disagree	2
E. Strongly disagree	1

9 Technology supports the rituals of the site (process).

A. Strongly agree 5

B. Agree 4

C. Neutral 3

D. Disagree 2

E. Strongly disagree 1

10 Technology supports the rites that in turn support the rituals.

A. Strongly agree 5

B. Agree 4

C. Neutral 3

D. Disagree 2

E. Strongly disagree 1

11 Rituals exist within the process to permit acquisition and use of new software without having to go through large amounts of red tape.

A. Strongly agree 5

B. Agree 4

C. Neutral 3

D. Disagree 2

E. Strongly disagree 1

12 Rituals exist to permit continuous upgrading of software products in order to stay current with the vendor's releases.

A. Strongly agree 5

B. Agree 4

C. Neutral 3

D. Disagree 2

E. Strongly disagree 1

13 Storytellers promote technology through their stories of successful implementations and the benefits delivered.

A. Strongly agree	5
B. Agree	4
C. Neutral	3
D. Disagree	2
E. Strongly disagree	1

14 Keepers of the Faith (mentors) promote the use of technology for sound reliability-based business decisions.

A. Strongly agree	5
B. Agree	4
C. Neutral	3
D. Disagree	2
E. Strongly disagree	1

15 Software and hardware issues are widely communicated to diminish the potential negatives that could be conveyed by the whisperers, gossips, and spies.

A. Strongly agree	5
B. Agree	4
C. Neutral	3
D. Disagree	2
E. Strongly disagree	1

16 There is a good communication plan to promote new technology through language and symbols.

A. Strongly agree	5
B. Agree	4
C. Neutral	3
D. Disagree	2
E. Strongly disagree	1

Exercise 13-5 The Technology Web – Analyzing the Low Spots

This exercise will help you use the technology web survey that you completed in Exercise 13-4 to address potentially-needed corrective actions within the four elements of culture.

Enter Your Technology Survey Scores.

Scores:

Category	Questions	Score	VP 1-4	P 5-8	N 9-12	G 13-16	VG 17-20
Organizational Values	1 to 4						
Role Models	5 to 8						
Rites & Rituals	9 to 12						
Cultural Infrastructure	13 to 16						

Enter your scores from the work process web survey in the "Score" column. Then mark an X in the appropriate column, based on the following key. Scores are classified as: VP = very poor (1 – 4 points); P = poor (5 – 8 points); A = average (9 – 12 points); G = good (13 – 16 points); and VG = very good (17 – 20 points)

2. Questions:

Which cultural scores are low (points between 1 and 8)?
Why do you believe that these cultural areas have scored this low?
From a technology perspective, what could be done to make positive improvement?
Which cultural scores are high (points between 13 and 20)?
Why do you believe that these cultural areas have scored this high?
From a technology perspective, what could you have done to generate improvements in the low scoring areas?

Chapter 14

Communication

Purpose

This chapter introduces the sixth of the eight elements of change—Communication. This critical element bonds people, groups, departments, and organization together with a common focus, keeping them aligned throughout the process of change.

What Is Included

Figure 14-1	The Communication Diagram
Exercise 14-1	What Piece of the Communication Process Is Missing?
Exercise 14-2	Receipt, Assumptions, and Feedback
Figure 14-2	Types of Communication Content
Exercise 14-3	Whispering Down the Lane
Exercise 14-4	Moments of High Influence (MoHi)
Exercise 14-5	Communication and the Four Elements of Culture
Exercise 14-6	The Communication Web of Change
Exercise 14-7	The Communication Web – Analyzing the Low Spots

Figure 14-1 The Communication Diagram

Part 1

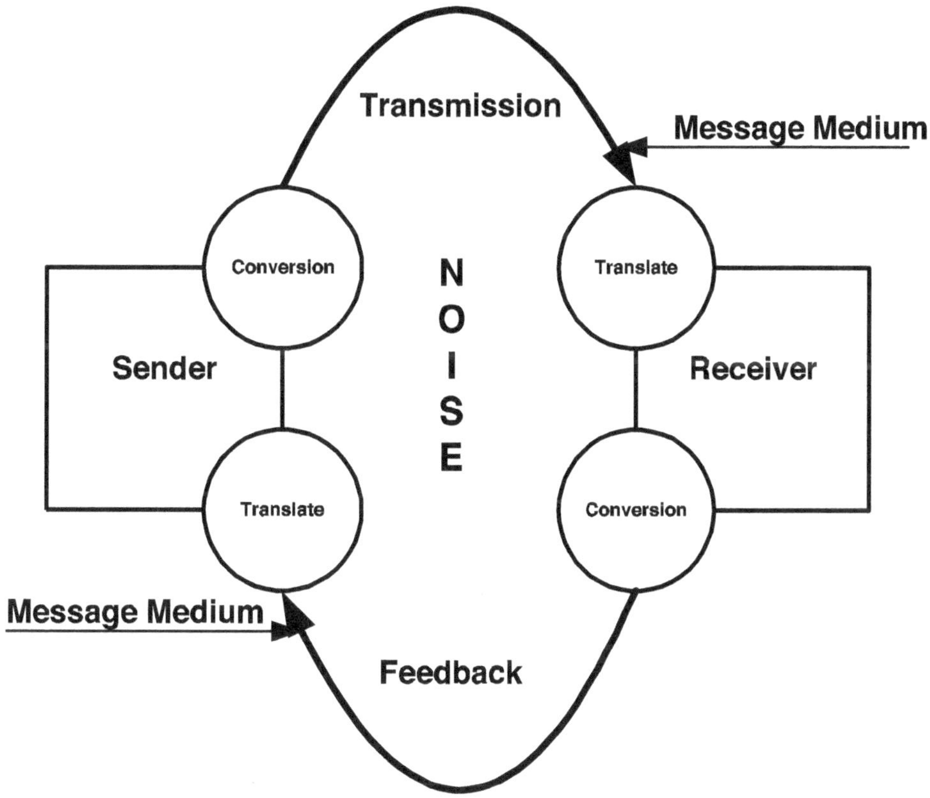

Figure 14-1 The Communication Diagram (continued)

Part 2

The Sender

Communication begins when, as the sender, you have a thought that you wish to communicate.

Conversion

You must now take your thought and convert it into a message that you can send to the person or group you want to receive it.

Transmission

Once your thought has been converted into a message, you need to send it.

Translation

When the message reaches the receiver, the message needs to be translated into something that the receiver can and will understand.

Receipt

The last part of the trip for the initial communication is its final receipt. This is the step when the receiver processes the message into action.

Feedback

The receiver has a responsibility to assure that what was sent was received. This check is done through the process of feedback.

Noise

Noise refers to everything that is external to the actual communication process, but interrupts or disrupts it.

Exercise 14-1 What Piece of the Communication Process Is Missing?

Figure 14-1 presented the communication diagram. In this exercise, review each scenario and indicate which part of the communication process was missed. There may be more than one part missed.

Scenario	Part Missed	Why?
Materials Management needed to alert Work Execution that delivery of a critical part was delayed. They sent an e-mail.		
They should have recognized that Work Execution didn't get the message when they got the call looking for the critical part.		
When the production line operator told the mechanic to clean the strainer on P-1A pump, the mechanic wasn't paying attention and proceeded to open the strainer on P-1B.		
The team missed the critical production meeting because the manager didn't tell them it had been scheduled for Monday at noon.		
Incorrect communication took place, resulting in a serious injury. The investigation found that mechanics discussed the job in a high-noise area; critical information was missed.		
When the engineering manager discussed the budget presentation, the subordinate thought that the manager wanted him to prepare the presentation. He did the work only to find the manager had already completed it.		
The reliability engineer wanted to alert the production foreman about a serious problem, but did not express himself well; the criticality of the problem was not understood. As a result, the line was shut down and production was lost.		

Exercise 14-2 Receipt, Assumptions, and Feedback

Most people know what they want to communicate. Often they are unsuccessful, resulting in miscommunication and people taking the wrong action. As the receiver, you can help to prevent this from taking place by making sure you receive the communication and by not making assumptions about what it means until you complete the feedback part of the process.

In this exercise, you will read each scenario, describe the assumptions that you might make, and then list how you could avoid taking the wrong action by employing the feedback process.

Communication Scenario	Initial Assumption	How You Would Get Feedback for Clarification
When you arrive at work, you are told about a leak in one of the feed lines to the production process. The entire maintenance force is being reassigned to the leak repair effort.		
The machinist foremen tell their work crews that new sophisticated laser a lignment equipment is being purchased; it will be used to replace the manual alignment method.		
Your planning team is informed that the upcoming major outage will be planned by contractors, not your group.		
You receive an e-mail from your boss informing you that you missed an important step in the project plan you developed. Every word of the e-mail is capitalized, in bold text and 18-point font.		
Your manager asks you to develop a presentation discussing the most recent project you completed.		

Figure 14-2 Types of Communication Content

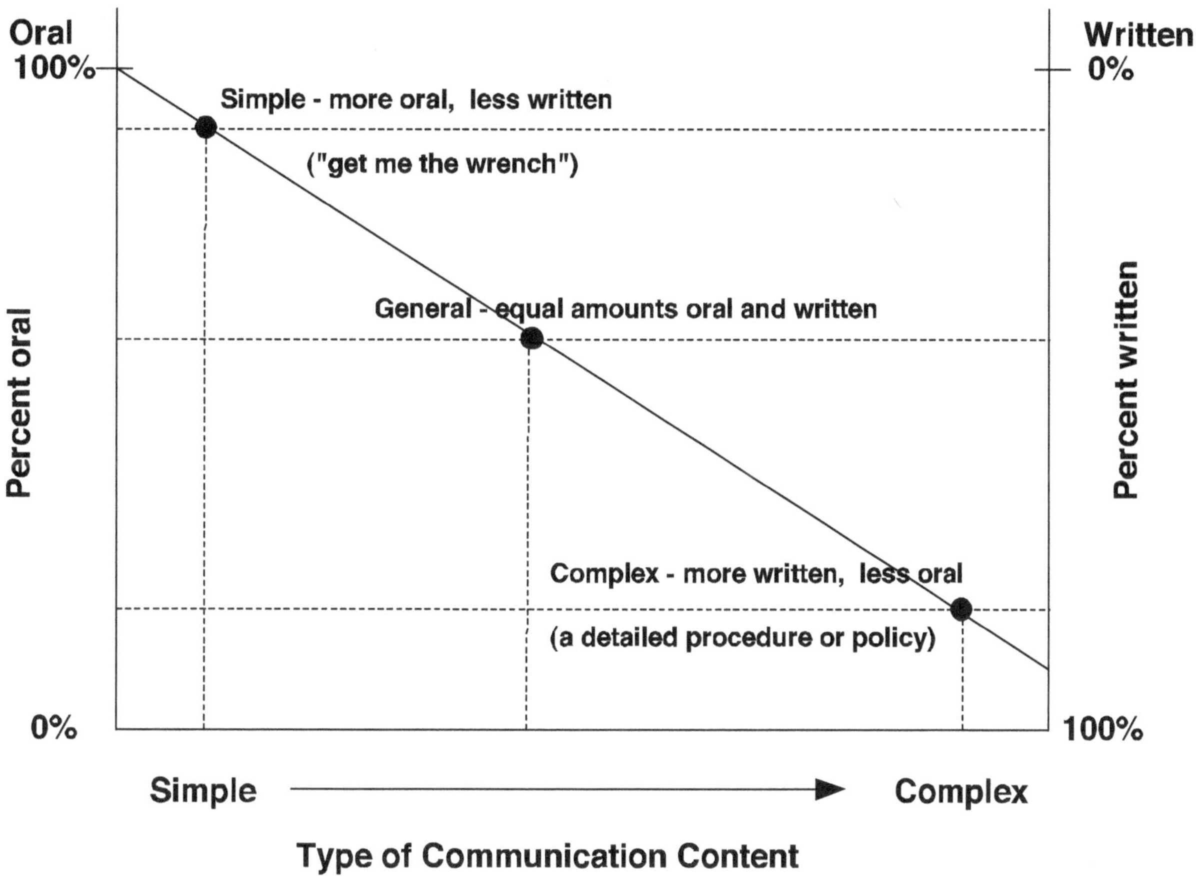

Exercise 14-3 Whispering Down the Lane

Part 1

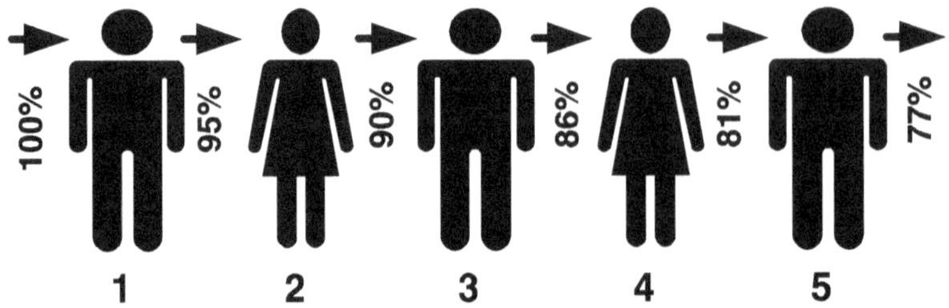

When we were young, we often played a party game called "whispering down the lane." Everyone sat in a line and the leader whispered a message in the ear of the first person. That person repeated the message to the second one in line and so on. The last person in line then repeated the information to the group who, in turn, were amused by the fact that the original message had virtually no resemblance to what was delivered at the end.

Why does this happen?

Exercise 14-3 Whispering Down the Lane

Part 2

The maintenance organization of Company X has decided to move from work crews assigned to specific areas within the plant to a single crew that is centrally dispatched to work on planned and scheduled jobs throughout the entire site. This change has been made by the maintenance manager in response to reliability problems that have increased throughout the production department. The manager believes this change would allow the correct resources to be applied to the most important jobs.

Using the diagram on the next page, describe how the message that the maintenance manager communicates to his staff and to the production manager could become distorted as it moves through the organization shown in the diagram that follows.

Describe how the message could be miscommunicated. Use your imagination.

Exercise 14-3 Whispering Down the Lane (continued)

Communication Diagram

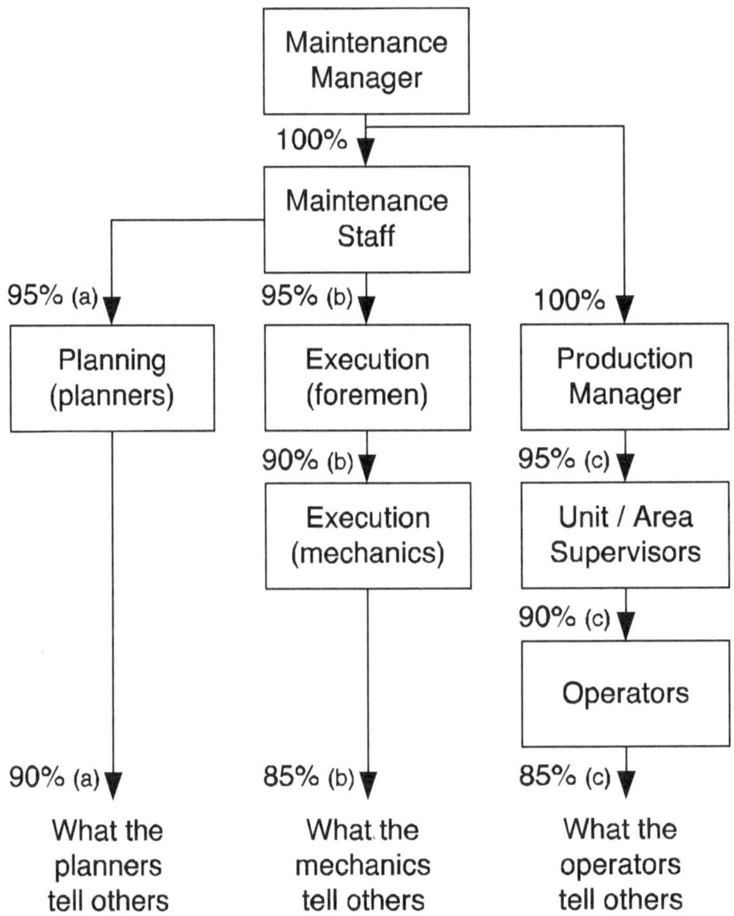

a, b, c each depict the original message but as it progressed through the different chains it ceased to be the same.

Exercise 14-3 Whispering Down the Lane

Part 3

Relate personal examples of this process in your plant or work group. What was the original communication? How did it come out at the other end of the communication chain and what could have been done to prevent the problem? Your examples do not have to be limited to oral communication. It can be any type of communication that gets distorted as it is passed from one person to another.

The original communication?
How did it come out at the other end of the communication chain?
What problems did this miscommunication cause?
How could these problems have been avoided?

Exercise 14-4 Moments of High Influence (MoHi)

In everyday life both in and out of work, there are times and places where people are in a high state of readiness to be influenced. Identify a MoHi event that has occurred in your plant. How was the communication process employed to derive value from it? If no benefit was derived, why did the organization miss this opportunity and what should have been done differently?

MoHi Event
What was the derived benefit?
How was the benefit achieved?
What was the missed benefit?
What should have been done differently?

Exercise 14-5 Communication and the Four Elements of Culture

1. **What is your plant's vision for its reliability and maintenance functions and how has communication been used to support it?**

2. **Organizational Values**

a. How have the organizational values been supported by your communication processes in order to achieve the vision?

b. Where are the gaps? What do you see as your opportunities for improvement?

c. What has been done to sustain / improve the communication process over time?

3. **Role Models**

a. How have the existing role models helped the organization communicate effectively?

a. If they have not provided proper communication, why not?

b. What do you see as your opportunities to improve communication by involving the existing or new role models? Specifically focus on plant reliability.

4. **Rites and Rituals**

a. How does communication affect the existing rites and rituals related to improved reliability and maintenance?

b. If you plan to change your organization's focus from reactive to proactive maintenance, how would you change the communication aspect of your rites and rituals?

5. **Cultural Infrastructure:** Describe how each of the roles within the cultural infrastructure contribute or fail to contribute to the communication process within your organization.

a. Story Tellers

b. Keepers of the Faith

c. Whisperers

d. Gossips

e. Spies

f. Symbols

g. Language

Exercise 14-6 The Communication Web of Change

The communication section of the web of change survey is divided into four parts, as shown below. By answering the survey's 16 questions and entering your scores on the four-axis web, you will see how Communication scores within the four elements of culture.

Scores:

Category	Questions	Score
Organizational Values	1 to 4	
Role Models	5 to 8	
Rites and Rituals	9 to 12	
Cultural Infrastructure	13 to 16	
Total Score – Communication	———	

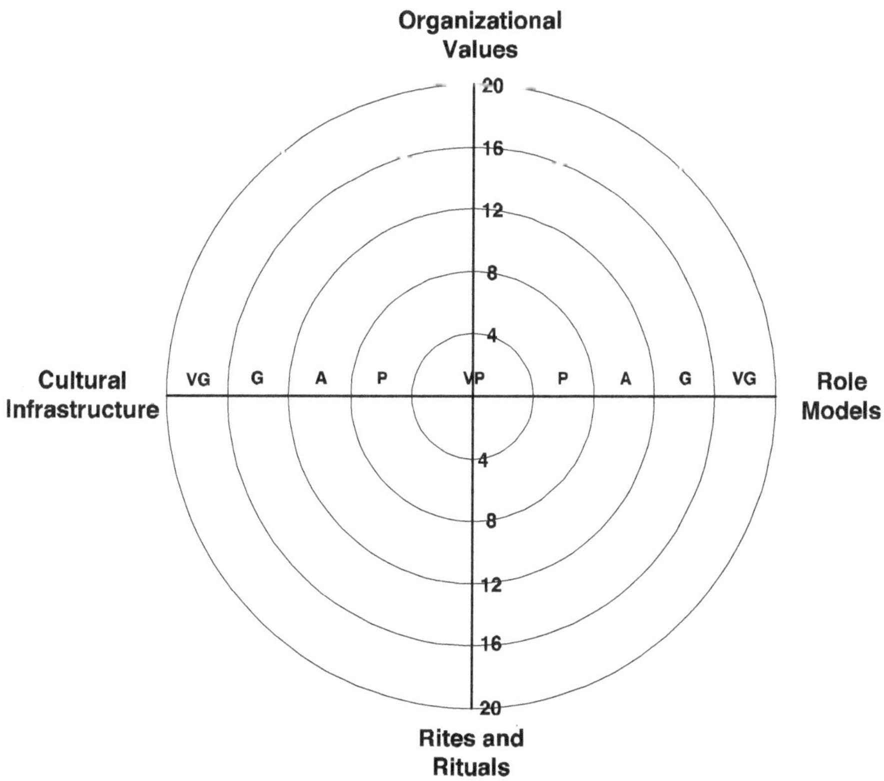

F. Communication

		Points
1	The organizational values are widely communicated.	
	A. Strongly agree	5
	B. Agree	4
	C. Neutral	3
	D. Disagree	2
	E. Strongly disagree	1
2	There are on-going processes in place to assure that the organizational values are understood.	
	A. Strongly agree	5
	B. Agree	4
	C. Neutral	3
	D. Disagree	2
	E. Strongly disagree	1
3	Where questions exist about the organizational values and their interpretation, feedback is provided for clarification.	
	A. Strongly agree	5
	B. Agree	4
	C. Neutral	3
	D. Disagree	2
	E. Strongly disagree	1
4	There is a continuous process in place to keep the workforce informed about the organizational values and how they should be applied in the day-to-day work.	
	A. Strongly agree	5
	B. Agree	4
	C. Neutral	3
	D. Disagree	2
	E. Strongly disagree	1

5 Role models are good communicators of the organizational values.

A. Strongly agree	5
B. Agree	4
C. Neutral	3
D. Disagree	2
E. Strongly disagree	1

6 Role models are good listeners. They receive feedback that could either valida te that the new process is understood or establish the need for clarification.

A. Strongly agree	5
B. Agree	4
C. Neutral	3
D. Disagree	2
E. Strongly disagree	1

7 Role models provide continuous communication, not just one time or with minimal frequency.

A. Strongly agree	5
B. Agree	4
C. Neutral	3
D. Disagree	2
E. Strongly disagree	1

8 Role models make themselves available to the organization to be good communicators. They view it as a key component of their work responsibilities.

A. Strongly agree	5
B. Agree	4
C. Neutral	3
D. Disagree	2
E. Strongly disagree	1

9 Rites are clearly communicated and there is a process to check understanding.

A. Strongly agree	5
B. Agree	4
C. Neutral	3
D. Disagree	2
E. Strongly disagree	1

10 Rituals are clearly communicated and there is a process to check understanding.

A. Strongly agree	5
B. Agree	4
C. Neutral	3
D. Disagree	2
E. Strongly disagree	1

11 Communication regarding rites and rituals has a lot of "why we are doing this and what are the benefits" content.

A. Strongly agree	5
B. Agree	4
C. Neutral	3
D. Disagree	2
E. Strongly disagree	1

12 Communication is widely used to reinforce the work process, the rituals, and their supporting rites.

A. Strongly agree	5
B. Agree	4
C. Neutral	3
D. Disagree	2
E. Strongly disagree	1

13 Storytellers' communication supports the process instead of telling how it was done around here before the change.

A. Strongly agree	5
B. Agree	4
C. Neutral	3
D. Disagree	2
E. Strongly disagree	1

14 Keepers of the Faith (mentors) focus their communication more on the new process instead of trying to show how the old way was better.

A. Strongly agree	5
B. Agree	4
C. Neutral	3
D. Disagree	2
E. Strongly disagree	1

15 A large amount of communication keeps everyone informed, diminishing the value of the gossips.

A. Strongly agree	5
B. Agree	4
C. Neutral	3
D. Disagree	2
E. Strongly disagree	1

16 There is a large amount of communication that keeps everyone informed, diminishing the value of the whisperers.

A. Strongly agree	5
B. Agree	4
C. Neutral	3
D. Disagree	2
E. Strongly disagree	1

2. Questions:

Which cultural scores are low (points between 1 and 8)?
Why do you believe that these cultural areas have scored this low?
From a communication perspective, what could be done to make positive improvement?
Which cultural scores are high (points between 13 and 20)?
Why do you believe that these cultural areas have scored this high?
From a communication perspective, what could be done to generate improvements in the low scoring areas?

Exercise 14-7 The Communication Web – Analyzing the Low Spots

This exercise will help you use the Communication web survey that you completed in Exercise 14-6 to address potentially-needed corrective actions within the four elements of culture.

Enter Your Technology Survey Scores.

Scores:

Category	Questions	Score	VP 1-4	P 5-8	N 9-12	G 13-16	VG 17-20
Organizational Values	1 to 4						
Role Models	5 to 8						
Rites & Rituals	9 to 12						
Cultural Infrastructure	13 to 16						

Enter your scores from the work process web survey in the "Score" column. Then mark an X in the appropriate column, based on the following key. Scores are classified as: VP = very poor (1 – 4 points); P – poor (5 – 8 points); A = average (9 – 12 points); G = good (13 – 16 points); and VG = very good (17 – 20 points)

Chapter 15

Interrelationships

Purpose

This chapter introduces the seventh of the eight elements of change—interrelationships. All of the other elements could be very effective, but without satisfactory interrelationships you still may have real problems.

What Is Included

Figure 15-1	The Definition of Interrelationships
Exercise 15-1	Your Existing Interrelationships
Figure 15-2	Vertical and Horizontal Interrelationships
Exercise 15-2	Reciprocity
Exercise 15-3	Building Interrelationships at ATPCo
Exercise 15-4	Building Your Own Interrelationships
Exercise 15-5	Interrelationships and the Other Seven Elements of Change
Exercise 15-6	Interrelationships and its Effect of the Four Elements of Culture
Exercise 15-7	The Interrelationships Web of Change
Exercise 15-8	The Interrelationships Web – Analyzing the Low Spots

Figure 15-1 The Definition of Interrelationships

Mutual and reciprocal positive feelings about the members of our organization, department, work team, or colleagues. A strong belief that other individuals with whom we interact are trustworthy and hold the goals of the group above their own.

Exercise 15-1 Your Existing Interrelationships

Describe a positive interrelationship that you have with someone in your organization. How does that interrelationship improve your ability as well as the ability of the work group to accomplish their initiatives?

Describe a negative interrelationship that you have with someone in your organization. How does that interrelationship make it difficult for you and your work group to accomplish their initiatives?

Figure 15-2 Vertical and Horizontal Interrelationships

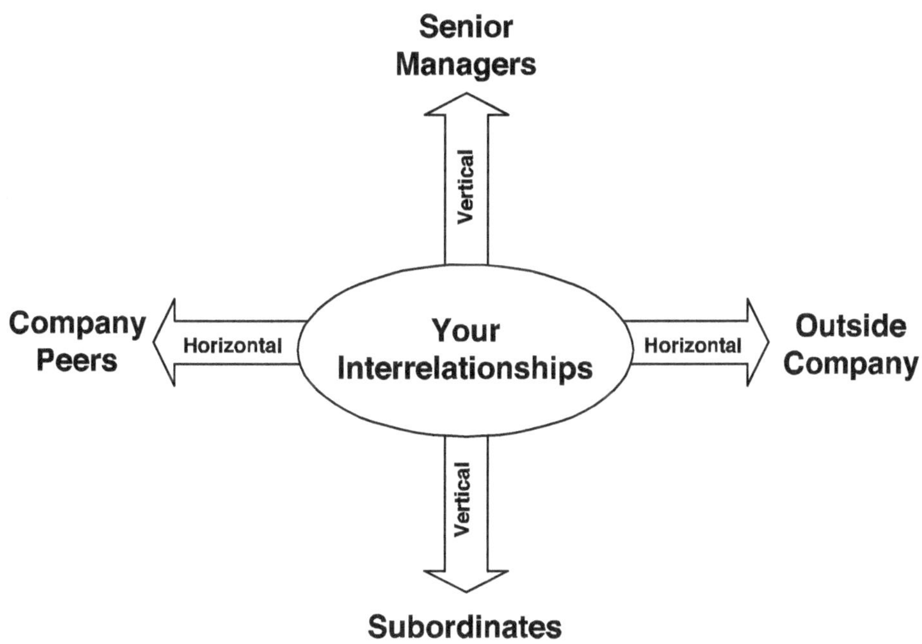

Exercise 15-2 Reciprocity

Reciprocity often has a negative context. Each of us has seen reciprocity at work. For example, a saleswoman may give us a free sample with the expectation that, by setting up a reciprocity arrangement, we will purchase her product. This emotional response receiving something and then feeling compelled to give something in returnis referred to as reciprocity.

However, it doesn't have to be negative. The concept of reciprocity can be utilized to improve interrelationships. Review the following scenarios. Then describe how reciprocity could generate a reciprocal positive reaction.

Scenario	Possible Reciprocal Results
1. Your manager needs a foreman to work the night shift on an upcoming outage. You volunteer.	
2. You attend a meeting and, without being asked, offer to take the notes.	
3. You recognize that a junior engineer is having a problem with a project. Even though you are very busy, you take time to help.	

Scenario	Possible Reciprocal Results
4. You recognize that supervisory training is lacking in your organization. Although it is not your job to provide it, you develop a simplified course to address this problem.	
5. Your work team has just completed a project. The conclusion that you have reached is very controversial. You offer to make the presentation.	
6. You receive a phone call from the mechanics working on the night shift. They believe that they are being asked to take on a job that is not safe. The on-shift production supervisor disagrees. You drive into the plant to resolve the problem.	

Exercise 15-3 Building Interrelationships at ATPCo

As the maintenance manager at ATPCo's Eastern Plant, you struggled to succeed in the plant when it was run by Mike Kane. Yet, you never really believed that his philosophy for the equipment–run it hard, fix it fast–was the optimum way to operate a production plant. With the introduction of the new product Epsilon, along with the associated increased production demands and the current level of poor equipment reliability from Mike's poor operating philosophy over the last ten years, you can clearly see that a new approach is going to be key to success.

With Mike's retirement, it is clear that it is time for a change. The new plant manager validates this because on his first day on site he asks that everyone focus their attention on making production quotas, but doing so with a focus on equipment reliability.

Describe how you would go about building strong interrelationships horizontally and vertically to accomplish the reliability vision of the new plant manager.

Exercise 15-4 Building Your Own Interrelationships

Describe several situations where you have built strong interrelationships (horizontally and vertically) within your own plant to accomplish your goals.

Exercise 15-5 Interrelationships and the Other Elements of Change

Scenarios for this Exercise: Seven scenarios are listed in this exercise. Identify which elements of change are affected by circling those elements and explain why.

Leadership	Work Process	Structure	Group Learning	Communication	Interrelationships	Rewards

Scenario: The wrong gasket was delivered to the job site. Each department production, maintenance, reliability, and materials management blamed the others, creating a strained work environment. The biggest issue was that the same problem happened the last time this same job was executed.

Why do the elements you marked relate to this scenario?

Leadership	Work Process	Structure	Group Learning	Communication	Interrelationships	Rewards

Scenario: The outside consultant hired to review how work gets accomplished and how improvements can be made has commented that the organization's inability to work across functional lines is a serious problem.

Why do the elements you marked relate to this scenario?

Leadership	Work Process	Structure	Group Learning	Communication	Interrelationships	Rewards

Scenario: Production and maintenance supervision are openly hostile to one another. Their hostility has nothing to do with the large backlog or the repair process. Apparently maintenance had discovered that production received a higher yearly salary increase because plant management wanted to recognize their contribution to improved production rates.

Why do the elements you marked relate to this scenario?

Leadership	Work Process	Structure	Group Learning	Communication	Interrelationships	Rewards

Scenario: Once again, maintenance and production are having problems. Production has been breaking more equipment lately, forcing maintenance to work excessive overtime to make the needed repairs. This has angered those in maintenance making the repairs. On the other hand, production has openly criticized maintenance for being too slow and not performing quality work. They are clearly not aligned.

Why do the elements you marked relate to this scenario?

Leadership	Work Process	Structure	Group Learning	Communication	Interrelationships	Rewards

Scenario: The reliability engineers are housed separately from their work teams. In fact, they are located on the opposite side of the plant, making day-to-day work relations difficult. Often they are not consulted on repair decisions for this reason.

Why do the elements you marked relate to this scenario?

Leadership	Work Process	Structure	Group Learning	Communication	Interrelationships	Rewards

Scenario: The relationship between maintenance and human resources was poor, but it has gotten worse due to the latest request to rewrite all of the maintenance job descriptions within the next month. The thing that angered the maintenance manager the most was that the request was made via e-mail.

Why do the elements you marked relate to this scenario?

Leadership	Work Process	Structure	Group Learning	Communication	Interrelationships	Rewards

Scenario: The reliability group was angry at systems because there had been a significant time delay getting their corrosion software installed. On the other hand, systems felt that they needed the extra time to make certain that the software would work within the company's software infrastructure.

Why do the elements you marked relate to this scenario?

Exercise 15-6 Interrelationships and its Effect on the Four Elements of Culture

1. **What is your plant's vision for its reliability and maintenance functions?**

2. **Organizational Values**

a. How have the organizational values been supported by the interrelationships within the plant?

b. What do you see as your opportunities for improving your interrelationships in order to better support the organizational values?

3. **Role Models**

a. How have the existing role models helped provide the organization with positive interrelationships?

b. If they have not supported positive interrelationships, why not?

c. What do you see as your opportunities for improving plant interrelationships to help the organization achieve its vision and apply the organizational values?

4. **Rites and Rituals**

a. What are the existing rites and rituals associated with how your organization addresses reliability and maintenance?

b. How do interrelationships affect successful performance?

c. If you plan to change your organization's focus from reactive to proactive maintenance, what would need to be changed in the area of plant interrelationships?

5. **Cultural Infrastructure:** Describe how each of the roles within the cultural infrastructure contributes to the interrelationships within the plant. Also show if they are positive or negative in nature.

a. Story Tellers

b. Keepers of the Faith

c. Whisperers

d. Gossips

e. Spies

f. Symbols

g. Language

Exercise 15-7 The Interrelationships Web of Change

The interrelationships section of the web of change survey is divided into four parts, as shown below. By answering the survey's 16 questions and entering your scores on the four-axis web, you will see how interrelationships scores within the four elements of culture.

Scores:

Category	Questions	Score
Organizational Values	1 to 4	
Role Models	5 to 8	
Rites and Rituals	9 to 12	
Cultural Infrastructure	13 to 16	
Total Score – Interrelationships	———	

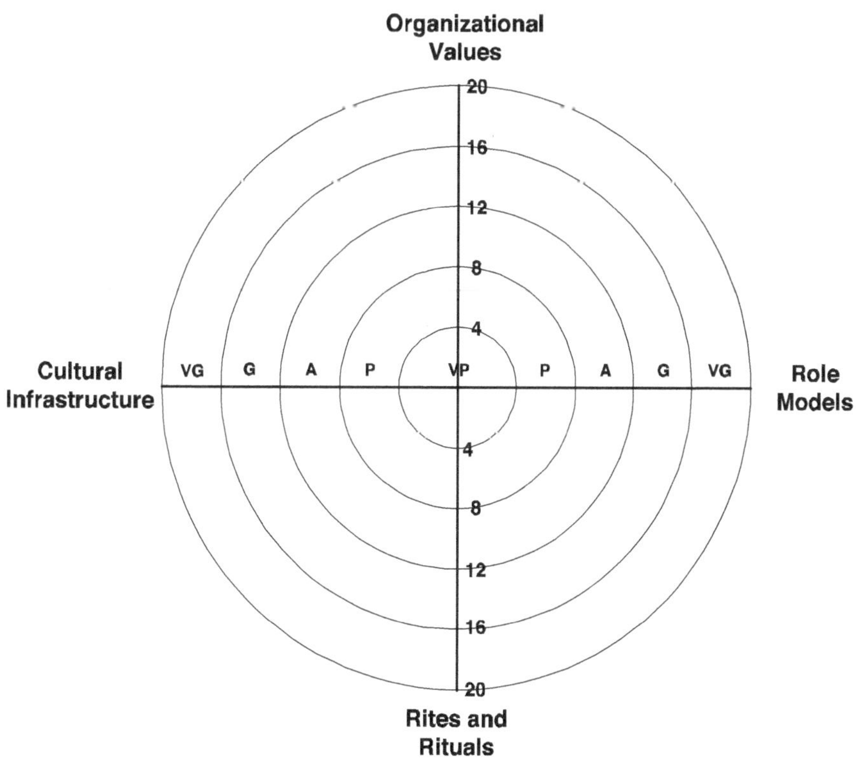

G. Interrelationships		Points
1	Organizational values are widely shared by the leadership team to promote buy-in and positive interrelationships.	
	A. Strongly agree	5
	B. Agree	4
	C. Neutral	3
	D. Disagree	2
	E. Strongly disagree	1
2	People at all levels work well together and are open to change.	
	A. Strongly agree	5
	B. Agree	4
	C. Neutral	3
	D. Disagree	2
	E. Strongly disagree	1
3	There is a single set of organizational values that is applied across all functions. There are no functional silos where the values apply differently dependent on your function.	
	A. Strongly agree	5
	B. Agree	4
	C. Neutral	3
	D. Disagree	2
	E. Strongly disagree	1
4	If audited, the site would show clear evidence that one set of values existed and was, in fact, being applied.	
	A. Strongly agree	5
	B. Agree	4
	C. Neutral	3
	D. Disagree	2
	E. Strongly disagree	1

5 Role models build organizational trust via positive interrelationships.

A. Strongly agree 5

B. Agree 4

C. Neutral 3

D. Disagree 2

E. Strongly disagree 1

6 Role models employ the concept of reciprocity in a positive manner.

A. Strongly agree 5

B. Agree 4

C. Neutral 3

D. Disagree 2

E. Strongly disagree 1

7 Role models create and support their allies in an effort to support the change process.

A. Strongly agree 5

B. Agree 4

C. Neutral 3

D. Disagree 2

E. Strongly disagree 1

8 Role models maintain both horizontal and vertical positive interrelationships.

A. Strongly agree 5

B. Agree 4

C. Neutral 3

D. Disagree 2

E. Strongly disagree 1

9	Horizontal interrelationships are a very important aspect of the site rituals.	
	A. Strongly agree	5
	B. Agree	4
	C. Neutral	3
	D. Disagree	2
	E. Strongly disagree	1
10	Vertical interrelationships are a very important aspect of the site rituals.	
	A. Strongly agree	5
	B. Agree	4
	C. Neutral	3
	D. Disagree	2
	E. Strongly disagree	1
11	Rituals that exist were built so that positive interrelationships are a result.	
	A. Strongly agree	5
	B. Agree	4
	C. Neutral	3
	D. Disagree	2
	E. Strongly disagree	1
12	Rites that exist (in support of rituals) were built so that positive interrelationships are promoted as a result.	
	A. Strongly agree	5
	B. Agree	4
	C. Neutral	3
	D. Disagree	2
	E. Strongly disagree	1

13 Storytellers promote positive interrelationships across the system by the stories that they relate.

A. Strongly agree	5
B. Agree	4
C. Neutral	3
D. Disagree	2
E. Strongly disagree	1

14 Keepers of the Faith (mentors) promote positive interrelationships as one of the best ways to get things done around here.

A. Strongly agree	5
B. Agree	4
C. Neutral	3
D. Disagree	2
E. Strongly disagree	1

15 Role models and other members of the management team promote positive interrelationships with the whisperers to enable themselves to get the real story about what is going on with the process.

A. Strongly agree	5
B. Agree	4
C. Neutral	3
D. Disagree	2
E. Strongly disagree	1

16 Role models and other members of the management team promote positive interrelationships with the gossips. This helps them find out what is going on with the process and, where necessary, correct miscommunications.

A. Strongly agree	5
B. Agree	4
C. Neutral	3
D. Disagree	2
E. Strongly disagree	1

Exercise 15-8 The Interrelationships Web – Analyzing the Low Spots

This exercise will help you use the interrelationships web survey you completed in Exercise 15-7 to address potentially-needed corrective actions within the four elements of culture.

1. Enter Your Interrelationships Survey Scores

Scores:

Category	Questions	Score	VP 1-4	P 5-8	N 9-12	G 13-16	VG 17-20
Organizational Values	1 to 4						
Role Models	5 to 8						
Rites & Rituals	9 to 12						
Cultural Infrastructure	13 to 16						

Enter your scores from the work process web survey in the "Score" column. Then mark an X in the appropriate column, based on the following key. Scores are classified as: VP = very poor (1 – 4 points); P = poor (5 – 8 points); A = average (9 – 12 points); G = good (13 – 16 points); and VG = very good (17 – 20 points)

2. Questions:

Which cultural scores are low (points between 1 and 8)?

Why do you believe that these cultural areas have scored this low?

From an interrelationships perspective, what could be done to make positive improvement?

Which cultural scores are high (points between 13 and 20)?

Why do you believe that these cultural areas have scored this high?

From an interrelationships perspective, what could be done to generate improvements in the low scoring areas?

Rewards

Chapter 16

Purpose

This chapter introduces the last of the eight elements of change—rewards. The other elements are integral to implementing the process of change in your plant. Rewards reinforce and support the efforts that are undertaken.

What Is Included

Figure 16-1	Definition
Figure 16-2	The Reactive Reward Process
Figure 16-3	The Proactive Reward Process
Exercise 16-1	Recognition of What Is Being Rewarded
Figure 16-4	Hierarchy of Needs
Exercise 16-2	Understanding What Fits Where in the Hierarchy
Exercise 16-3	Using the Hierarchy of Needs / Rewards to Motivate Change
Exercise 16-4	Negative and Neutral Rewards
Exercise 16-5	Rewards and the Four Elements of Culture
Exercise 16-6	The Rewards Web of Change
Exercise 16-7	The Rewards Web – Analyzing the Low Spots

Figure 16-1 Definition

Rewards...

The return for the performance of a behavior that is desired; a positive reinforcement.

Figure 16-2 The Reactive Reward Process

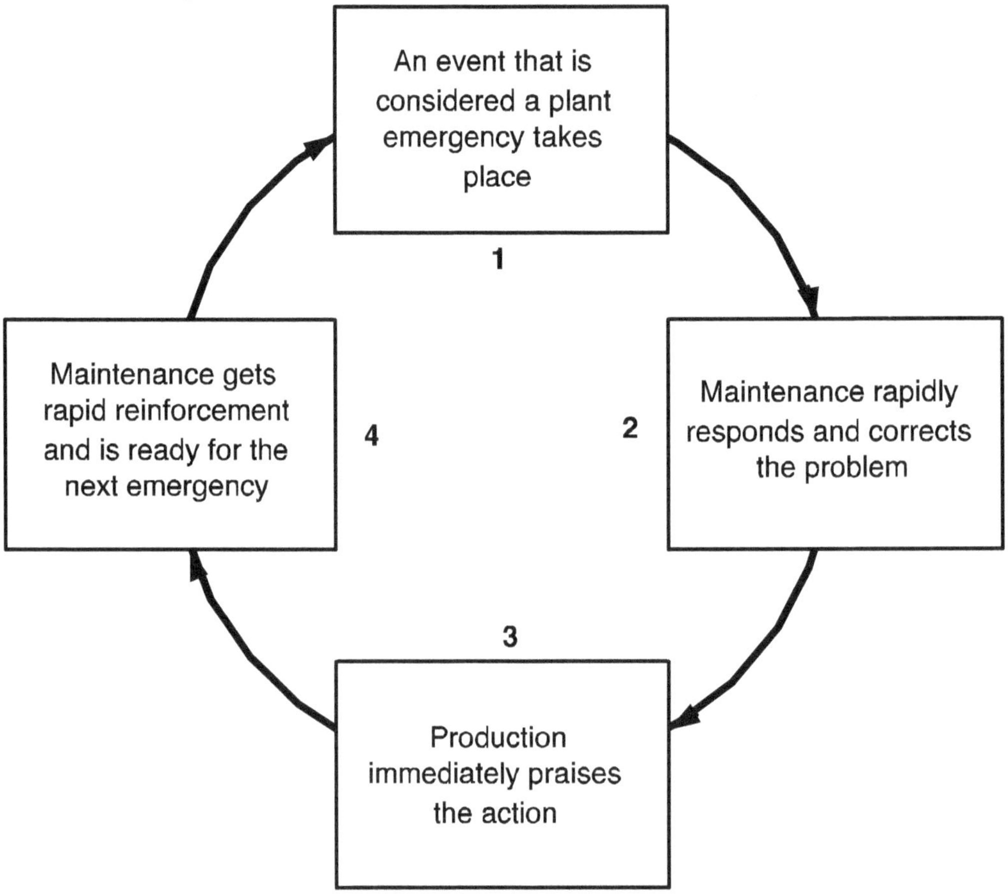

Figure 16-3 The Proactive Reward Process

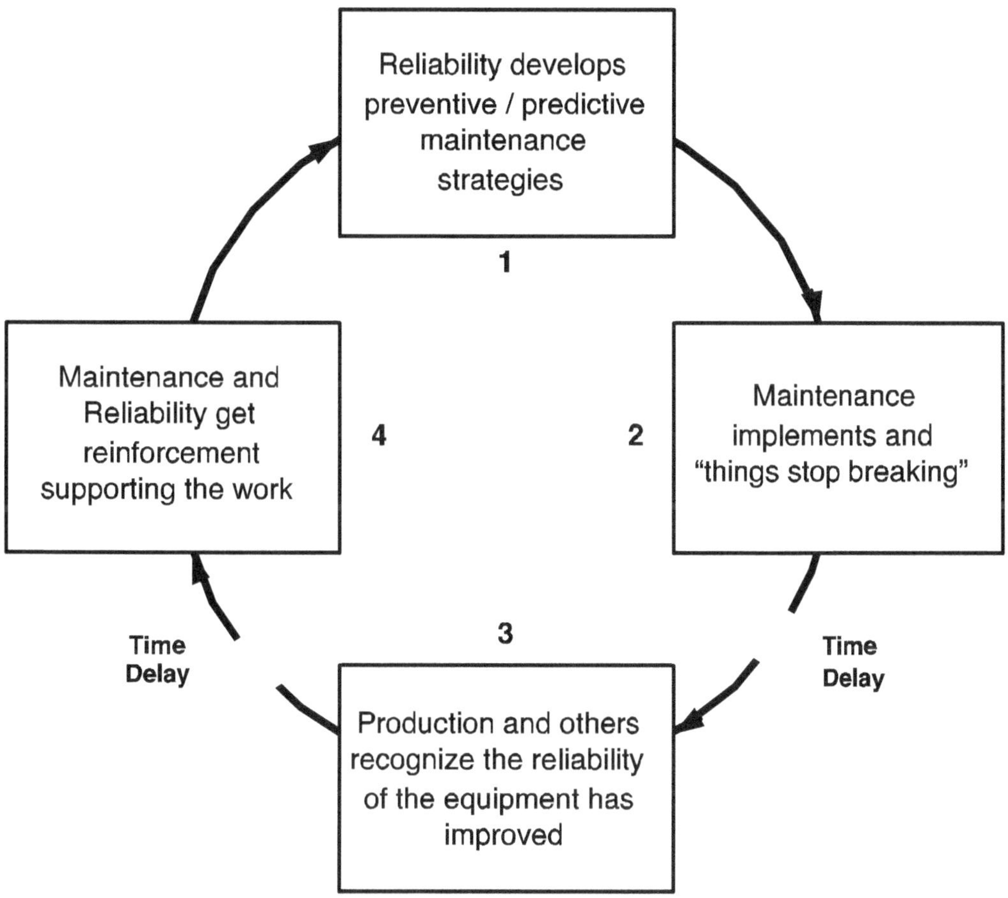

Exercise 16-1 Recognition of What Is Being Rewarded

Based on each scenario, determine if the reward being provided is for reactive or proactive work results. Place an "X" in the correct column.

Scenario	Reactive	Proactive
The field execution superintendent received a promotion for his efforts to identify the problems from the night shift and then to respond to them first thing the next day.		
Production praised the work crew that stopped what they were doing and responded to the plant problem, even though the problem wasn't as bad as production had portrayed it.		
The field foreman was thanked by his supervisor for refusing to respond to production's request to move his crew to a job that was planned for the next week simply because the production supervisor wanted it done today.		
Recognition was given to the planning group tha t created a set of work plans for the upcoming outage. The plans included input from all involved groups.		
A reward was given to the materials coordinator for expediting materials for a job that production said was important–even though the part was never picked up after delivery.		
Recognition was provided for the engineer who determined that an upgraded part would improve the mean time between failure for a critical plant asset. The engineer also got the execution team to install it at the next planned repair.		

Figure 16-4 Hierarchy of Needs

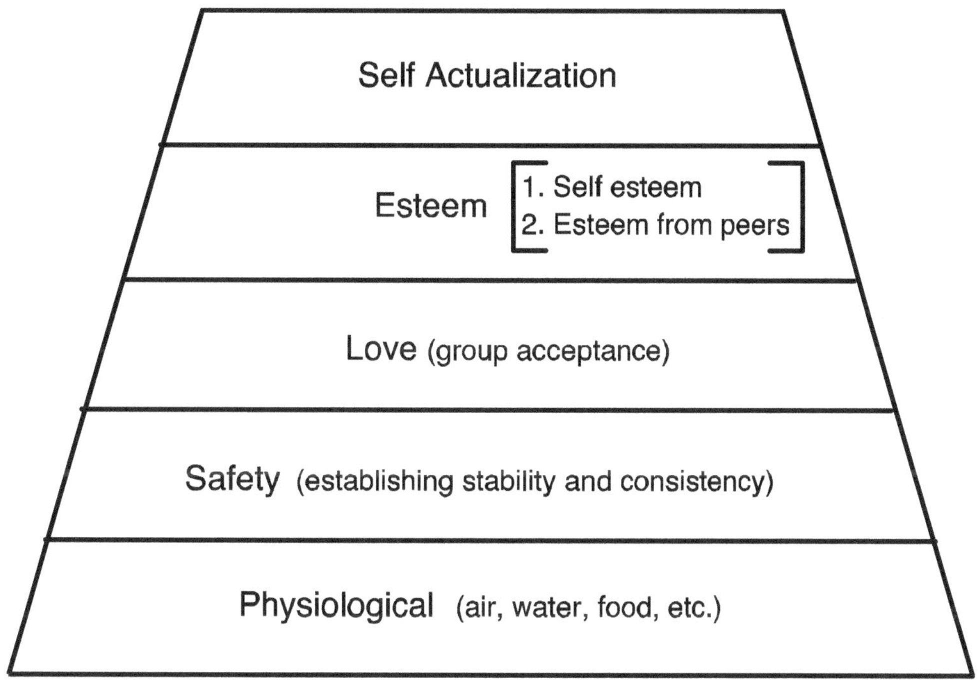

Exercise 16-2 Understanding What Fits Where in the Hierarchy

Match the statement to the correct level of the Hierarchy of Needs.

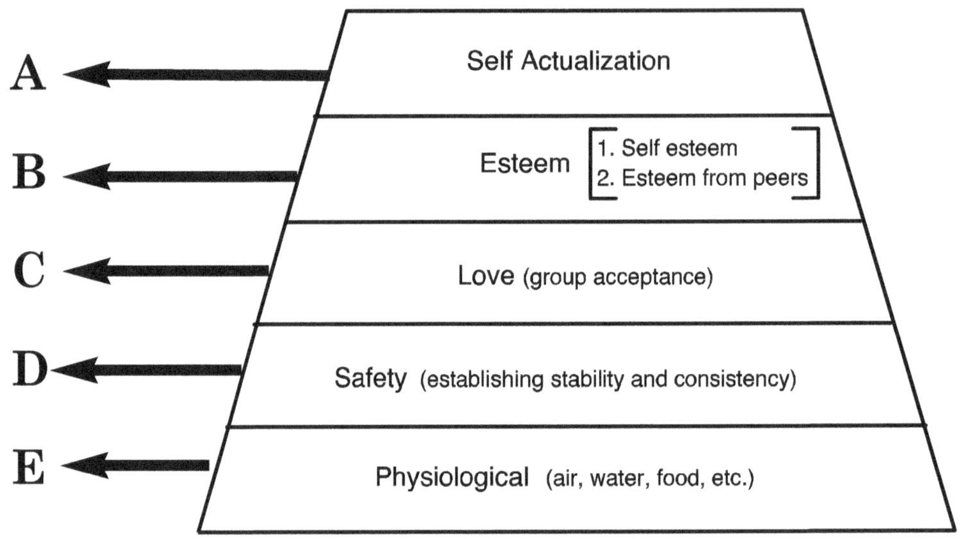

Statement	Match to A, B, C, D or E
Having money in the bank that will allow me to retire whenever I wish is very comforting.	
Being part of the Epsilon production team is a rewarding experience. We work together to get the job done	
When I finished the report and had my work accepted by management, I felt very good.	
The volunteer work that I do in my community is very rewarding and gives me a sense of accomplishment.	
I once worked as part of a maintenance team where I knew without a doubt that the other members would support our group–no matter what.	
We need food to survive.	
After my presentation to the senior leadership of the company, my peers told me that I had done a wonderful job.	
The "no layoff" clause in the union contract made the workforce feel secure.	
We need air to breath.	
The successful businessman retired and went back to school for his degree in anthropology–something he had always wanted to do.	

Exercise 16-3 Using the Hierarchy of Needs / Rewards to Motivate Change

The Hierarchy of Needs can be utilized to motivate change. You need to understand the level of the hierarchy that employees are currently on and then provide rewards that will help them move up the hierarchy. In the scenario provided, a change is desired for the plant, but it doesn't appear easy to implement. Determine how the Hierarchy of Needs could be used as a tool to help implement the change.

Scenario:

The plant has been having serious equipment problems. Production equipment is continually breaking down, maintenance can't seem to fix it soon enough, and no reliability-based improvements have been added even though they have been identified.

Production wants equipment repairs made as quickly as possible. Getting equipment fixed quickly helps them make their weekly production quotas even though the equipment doesn't stay in service very long. Maintenance is always rushed, causing mistakes and reworks. The reliability engineers are never told when the equipment is going in for repair so that they can have their improved reliability fixes implemented.

As the maintenance manager, you are determined to correct these problems. Your vision is that of planned work executed on a set schedule with more than enough time to get the reliability improvements implemented.

(continued)

Exercise 16-3 (continued)

How could you use the Safety need to convince people of the need to change?
How could you use the Love (group acceptance) need to promote the change initiative?
How could you use the Esteem (self or peers) need to implement your new work process?
How do you see Self Actualization fitting into the change process?

Exercise 16-4 Negative and Neutral Rewards

Not all rewards are positive. Some have a negative effect, whereas others are neutral and have no effect at all. Because the reward system is built to motivate organizations and people, these types of rewards are not desirable in any context.

Describe a time when you rewarded an individual or group and the reward turned out to be negative or neutral.
What was the positive outcome you sought when you conceptualized the reward?
What was the actual outcome?
How would you approach this reward situation in the future to avoid negative / neutral outcomes?

Exercise 16-5 Rewards and the Four Elements of Culture

1. **What is your plant's vision for its reliability and maintenance functions and how rewards have been used to support it?**

2. **Organizational Values**
   a. How have the organizational values been supported by your reward systems?

   b. Where are the gaps? What do you see as your opportunities for improvement?

   c. What has been done to sustain / improve the reward process over time?

3. **Role Models**
   a. How have the existing role models helped the organization reward the correct behaviors with a focus on the vision?

   b. If they have not provided proper rewards, why not?

   c. Focusing on reliability and maintenance, where are your opportunities for improvement of your reward system? How can the role models support these changes?

4. Rites and Rituals

a. What are the rewards associated with existing rites and rituals in order to address improved reliability and maintenance?

b. If you plan to change your organization's focus from reactive to proactive maintenance, how would you change the rites and rituals and their supporting rewards?

5. Cultural Infrastructure: Describe how each of the roles within the cultural infrastructure contribute or fail to contribute to the reward process within your organization.

a. Story Tellers

b. Keepers of the Faith

c. Whisperers

d. Gossips

e. Spies

f. Symbols

g. Language

Exercise 16-6 The Rewards Web of Change

The rewards section of the web of change survey is divided into four parts, as shown below. By answering the survey's 16 questions and entering your scores on the four-axis web, you will be able to see how rewards scores within the four elements of culture.

Scores:

Category	Questions	Score
Organizational Values	1 to 4	
Role Models	5 to 8	
Rites and Rituals	9 to 12	
Cultural Infrastructure	13 to 16	
Total Score – Interrelationships	———	

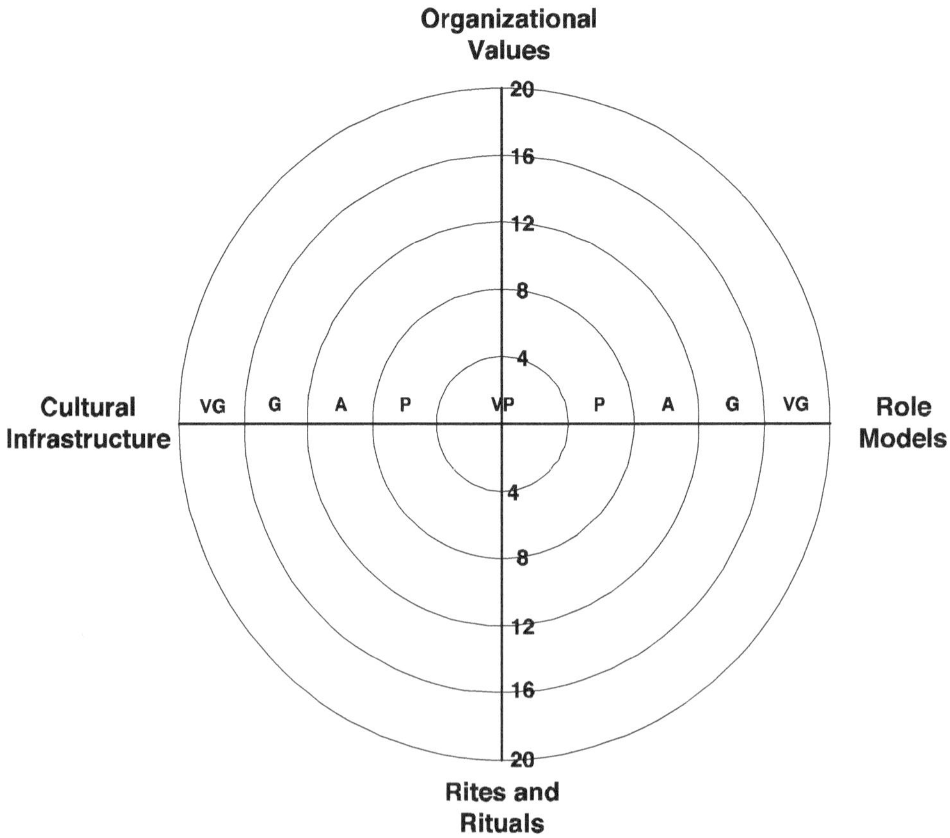

H. Rewards **Points**

1 The organizational values include a reward process that supports its success.

A. Strongly agree 5

B. Agree 4

C. Neutral 3

D. Disagree 2

E. Strongly disagree 1

2 The organization clearly understands the reward process as well as how it is linked to the organizational values.

A. Strongly agree 5

B. Agree 4

C. Neutral 3

D. Disagree 2

E. Strongly disagree 1

3 The Goal Achievement Model (or a similar process) is used to promote the values through the vision-goals-initiative-activity process.

A. Strongly agree 5

B. Agree 4

C. Neutral 3

D. Disagree 2

E. Strongly disagree 1

4 The system provides negative or neutral rewards for courses of action that do not support the value system being utilized.

A. Strongly agree 5

B. Agree 4

C. Neutral 3

D. Disagree 2

E. Strongly disagree 1

5 The management team recognizes that money is not a long-term motivational reward for work well done. Alternative rewards are provided.

A. Strongly agree	5
B. Agree	4
C. Neutral	3
D. Disagree	2
E. Strongly disagree	1

6 The time delay between successful action and its reward is minimized wherever possible

A. Strongly agree	5
B. Agree	4
C. Neutral	3
D. Disagree	2
E. Strongly disagree	1

7 Role models who promote the work process and other change efforts are visibly rewarded by the organization so that others can see what success looks like.

A. Strongly agree	5
B. Agree	4
C. Neutral	3
D. Disagree	2
E. Strongly disagree	1

8 Role models who do not promote the change (or may even work against it) visibly receive neutral or negative rewards so that the workforce can see what failure looks like.

A. Strongly agree	5
B. Agree	4
C. Neutral	3
D. Disagree	2
E. Strongly disagree	1

9 Rituals (where appropriate) carry timely rewards that reinforce the change effort.

A. Strongly agree 5

B. Agree 4

C. Neutral 3

D. Disagree 2

E. Strongly disagree 1

10 Rites are designed to provide rewards that reinforce the rituals and the process.

A. Strongly agree 5

B. Agree 4

C. Neutral 3

D. Disagree 2

E. Strongly disagree 1

11 Where rewards do not reinforce the change effort, they are altered.

A. Strongly agree 5

B. Agree 4

C. Neutral 3

D. Disagree 2

E. Strongly disagree 1

12 Where there is a time delay that will diminish the value of a reward, an active effort is made to eliminate the delay or make it as small as possible.

A. Strongly agree 5

B. Agree 4

C. Neutral 3

D. Disagree 2

E. Strongly disagree 1

13 Members of the cultural infrastructure recognize that one reward if they remain in roles that are key to the infrastructure is their support for the change effort.

A. Strongly agree	5
B. Agree	4
C. Neutral	3
D. Disagree	2
E. Strongly disagree	1

14 Management actively engages and rewards cultural infrastructure members for their help in promoting the change.

A. Strongly agree	5
B. Agree	4
C. Neutral	3
D. Disagree	2
E. Strongly disagree	1

15 Negative and / or neutral rewards are provided to those members of the cultural infrastructure who use their roles to disrupt the change process.

A. Strongly agree	5
B. Agree	4
C. Neutral	3
D. Disagree	2
E. Strongly disagree	1

16 Spies are not rewarded for passing sensitive information. At the minimum, they are informed that they are disrupting the process.

A. Strongly agree	5
B. Agree	4
C. Neutral	3
D. Disagree	2
E. Strongly disagree	1

Exercise 16-7 The Rewards Web – Analyzing the Low Spots

This exercise will help you use the rewards web survey that you completed in Exercise 16-6 to address potentially-needed corrective actions within the four elements of culture.

Scores:

Category	Questions	Score	VP 1-4	P 5-8	N 9-12	G 13-16	VG 17-20
Organizational Values	1 to 4						
Role Models	5 to 8						
Rites & Rituals	9 to 12						
Cultural Infrastructure	13 to 16						

Enter your scores from the work process web survey in the "Score" column. Then mark an X in the appropriate column, based on the following key. Scores are classified as: VP = very poor (1 – 4 points); P = poor (5 – 8 points); A = average (9 – 12 points); G = good (13 – 16 points); and VG = very good (17 – 20 points)

2. Questions:

Which cultural scores are low (points between 1 and 8)?
Why do you believe that these cultural areas have scored this low?
From a rewards perspective, what could be done to make positive improvement?
Which cultural scores are high (points between 13 and 20)?
Why do you believe that these cultural areas have scored this high?
From a rewards perspective, what could be done to improve the low scoring areas?

Chapter 17

The Web of Cultural Change

Purpose

This chapter introduces you to the Web of Cultural Change. Throughout the workbook, you have taken surveys that enabled you to develop webs examining each of the elements of culture related to the eight elements of change and each of the eight elements of change related to the four elements of culture. Now it is time to put all of these together into a single web—the Web of Cultural Change.

What Is Included

Figure 17-1 What Is the Cultural Web of Change?

Figure 17-2 A Sample Web of Cultural Change

Figure 17-3 The Other Diagrams

Figure 17-4 A Web Example (Figure 17-8 from the Book)

Exercise 17-1 What Is the Web of Cultural Change Telling You?

Figure 17-5 The Sub Webs from Figure 17-4

Exercise 17-2 What are the Sub Webs of Cultural Change Telling You?

Figure 17-6 The Web Surveys We Have Done

Exercise 17-3 Your Web

Exercise 17-4 What Are Your Web Diagrams Telling You?

Figure 17-1 What Is the Cultural Web of Change?

The Web of Cultural Change is a tool that portrays an organization's cultural profile in term of the eight elements of change (the radial spokes of the diagram) and the four elements of culture (the web scores). With this information, an organization can analyze what they are or are not doing and take corrective action where required.

Figure 17-2 A Sample Web of Cultural Change

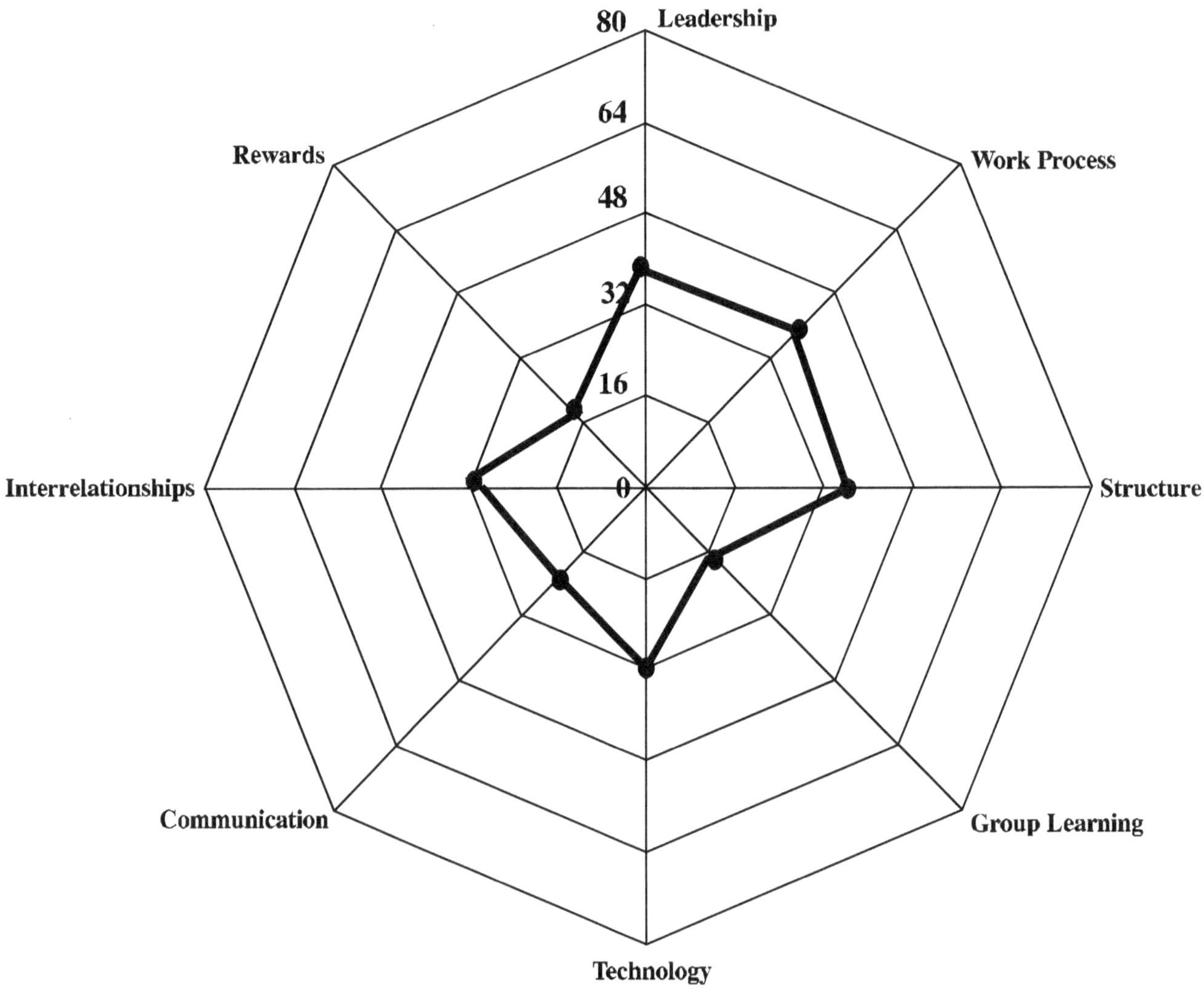

Figure 17-3 The Other Diagrams

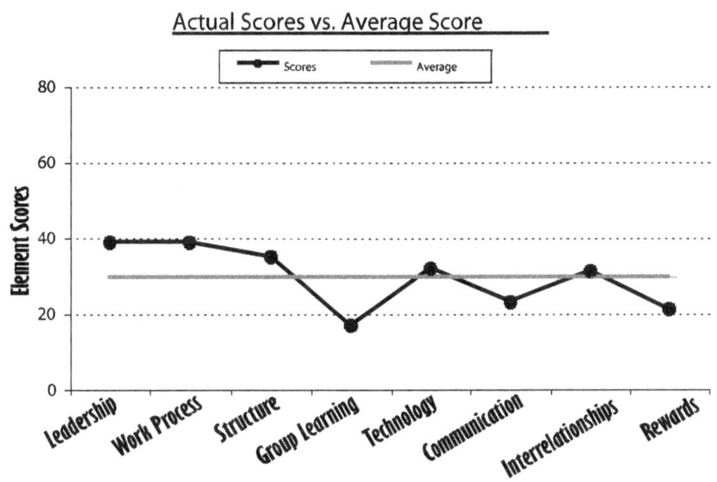

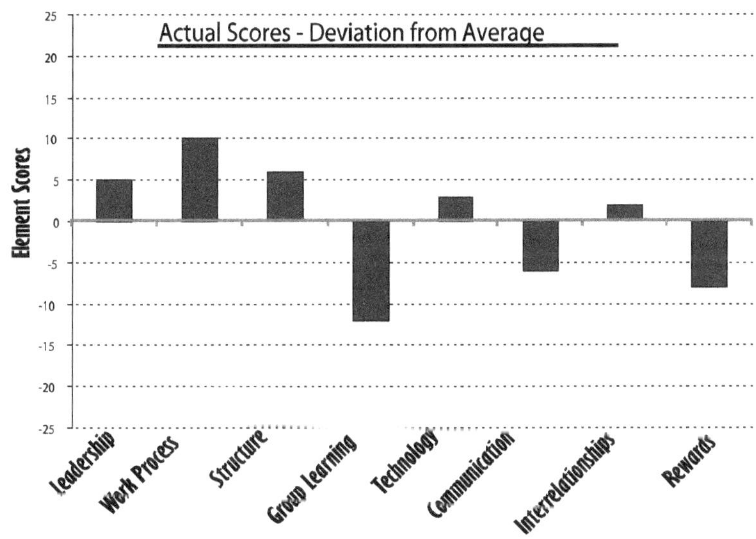

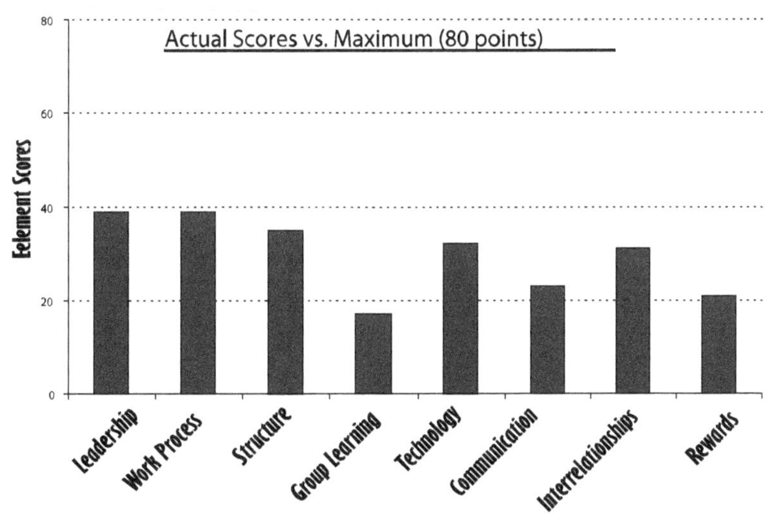

Figure 17-4 A Web Example (Figure 17-8 from the Book)

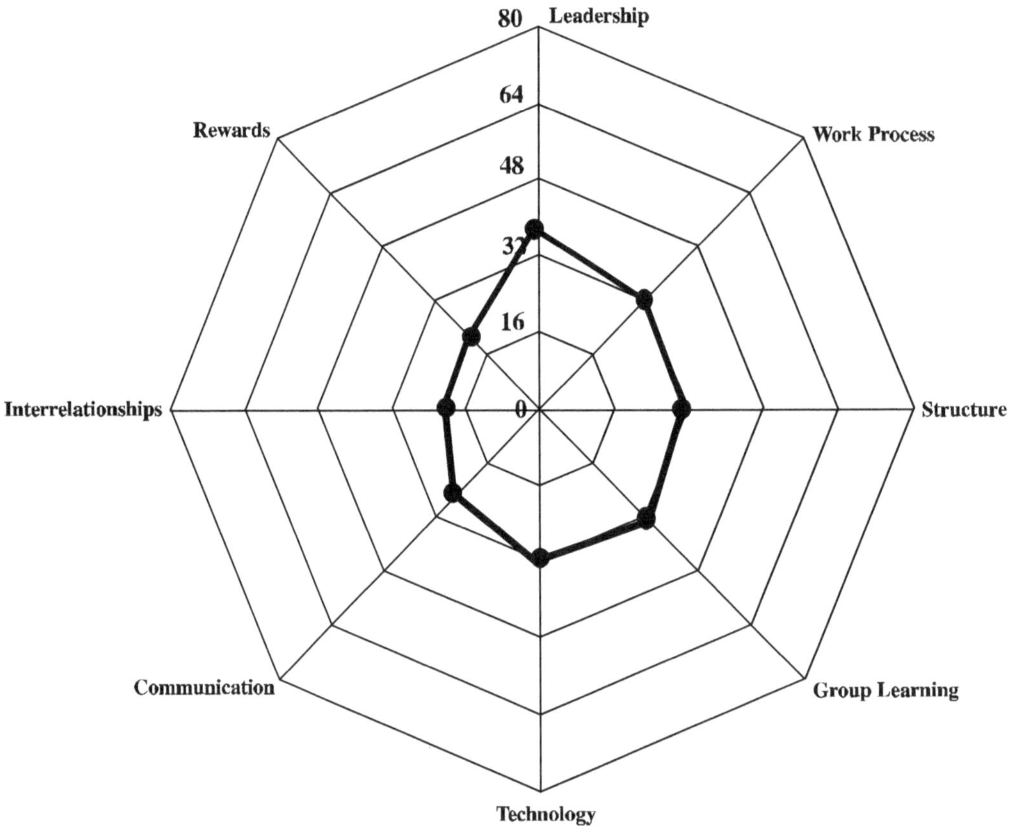

	Max Score	Baseline Score	Baseline vs Average
Leadership	80	38	9.00
Work Process	80	30	1.00
Structure	80	33	4.00
Group Learning	80	33	4.00
Technology	80	31	2.00
Communication	80	26	-3.00
Interrelationships	80	20	-9.00
Rewards	80	21	-8.00
Average		29.00	

The baseline scores shown on the table are the scores that are charted on the web diagram above.

Exercise 17-1 What Is the Web of Cultural Change Telling You?

Taking the Web of Cultural Change survey and charting your scores is only the start. You need to evaluate what the diagram is telling you and then, with your team, conduct further analysis to determine the root causes of the low scores and your problems. Looking at the chart and table in Figure 17-4, determine at a high level what the diagram is saying.

Web Analysis

Scores: (0 to 16 very poor, 17 to 32 poor, 33 to 48 average, 49 to 64 good and 65 to 80 very good)

What was the average score and in what range did it fall?
Which of the eight Elements of Change scored poor or very poor?
Which elements scored in the average range?
Which elements scored good or very good?
How could the poor and very poor scored elements cause problems with your change process?

Figure 17-5 The Sub Webs from Figure 17-4

Organizational Values

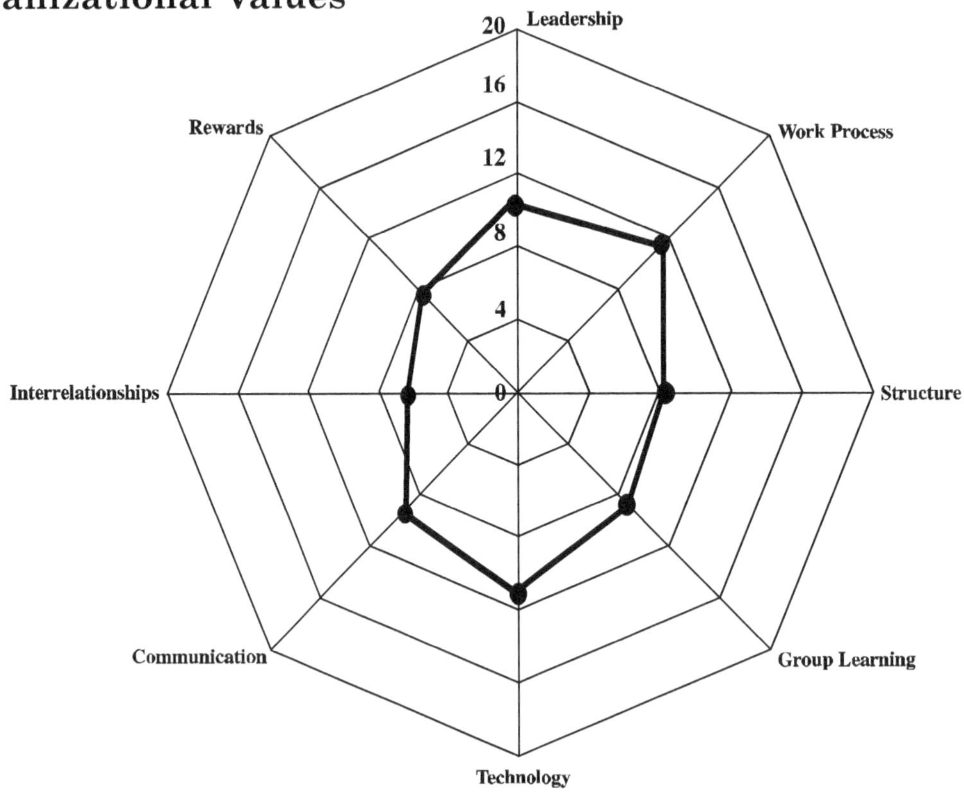

Role Models

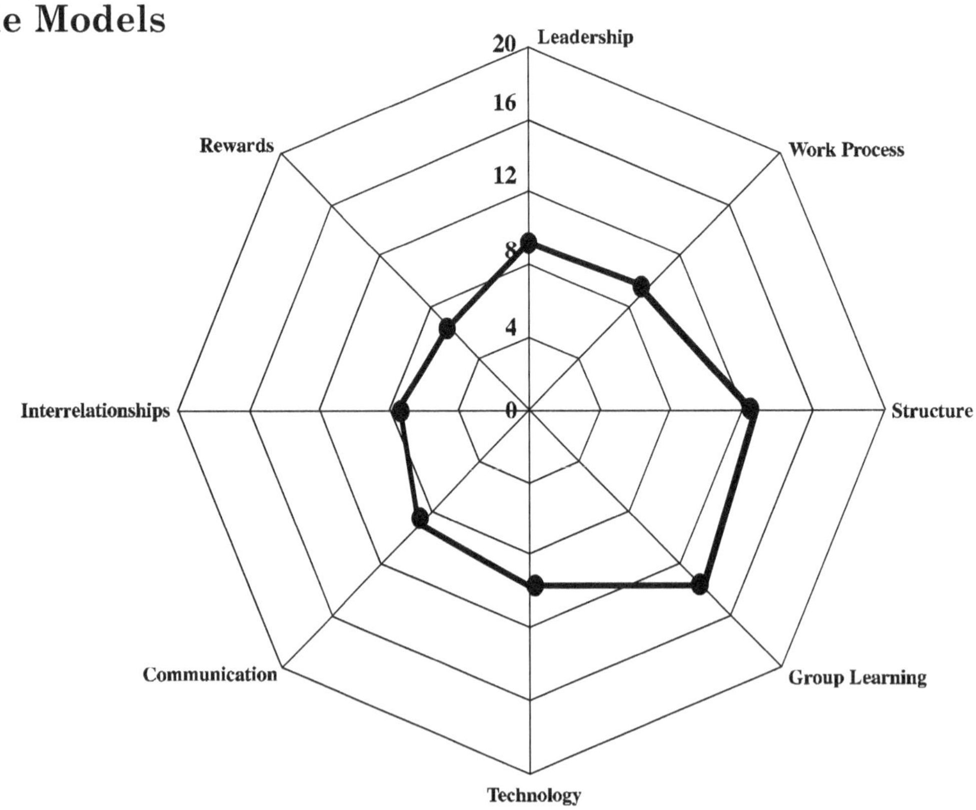

Figure 17-5 (continued)

Rituals / Rites

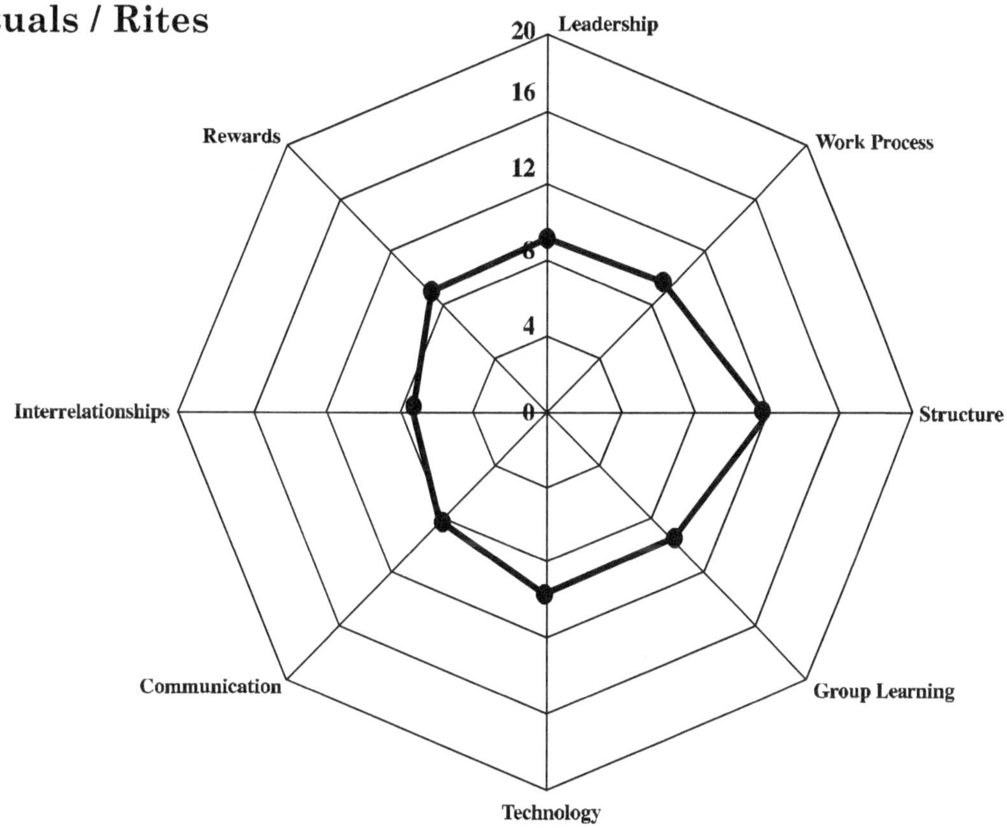

Cultural Infrastructure

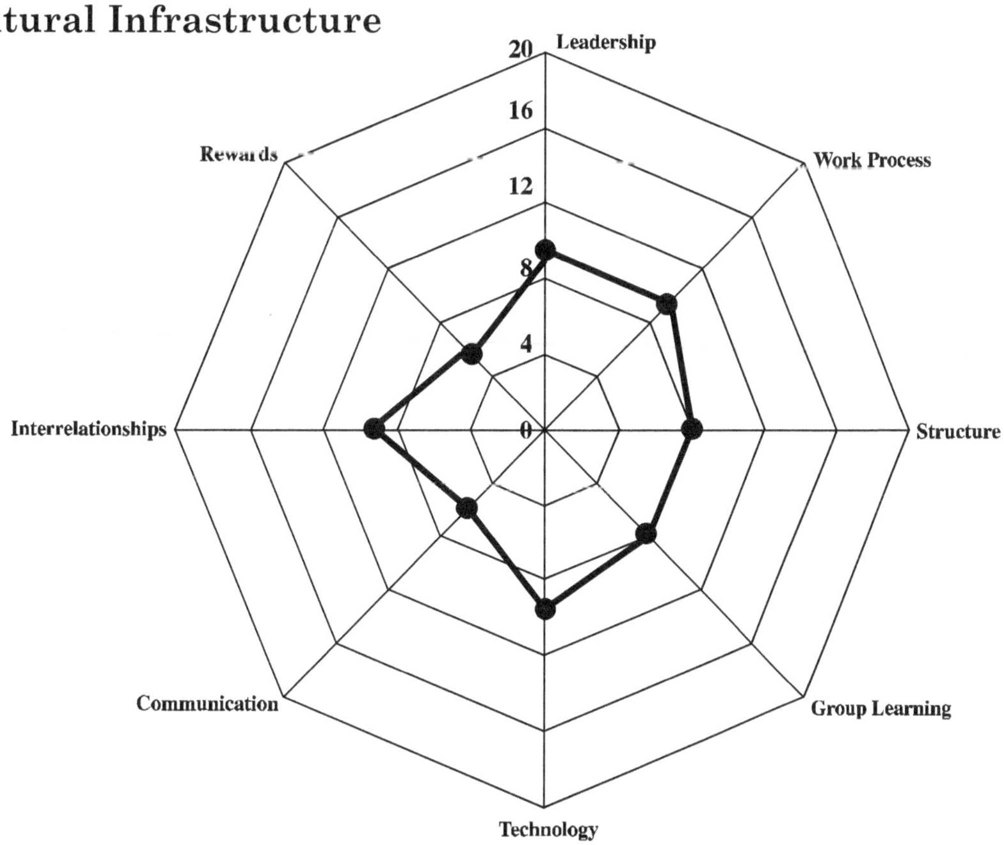

Exercise 17-2 What are the Sub Webs of Cultural Change Telling You?

The four figures shown in Figure 17-5 show the parts of the Web of Cultural Change from Figure 17-4. Not only can you learn something from the total web, but you can also learn something from its component parts: organizational values, roles models, rites and rituals, and cultural infrastructure.

Organizational Values

Scores: (0 to 4 very poor, 5 to 8 poor, 9 to 12 average, 13 to 16 good and 17 to 20 very good)

Which of the eight elements of change scored poor or very poor?
Which elements scored average?
Which elements scored good or very good?
How could the poor and very poor scored elements cause problems with your change process?

Roles Models

Scores: (0 to 4 very poor, 5 to 8 poor, 9 to 12 average, 13 to 16 good and 17 to 20 very good)

Which of the eight elements of change scored poor or very poor?
Which elements scored average?
Which elements scored good or very good?
How could the poor and very poor scored elements cause problems with your change process?

Rites and Rituals

Scores: (0 to 4 very poor, 5 to 8 poor, 9 to 12 average, 13 to 16 good and 17 to 20 very good)

Which of the eight elements of change scored poor or very poor?
Which elements scored average?
Which elements scored good or very good?
How could the poor and very poor scored elements cause problems with your change process?

Cultural Infrastructure

Scores: (0 to 4 very poor, 5 to 8 poor, 9 to 12 average, 13 to 16 good and 17 to 20 very good)

Which of the eight elements of change scored poor or very poor?
Which elements scored average?
Which elements scored good or very good?
How could the poor and very poor scored elements cause problems with your change process?

Figure 17-6 The Web Surveys We Have Done

This diagram shows the sources of web data from the various chapters of the workbook.

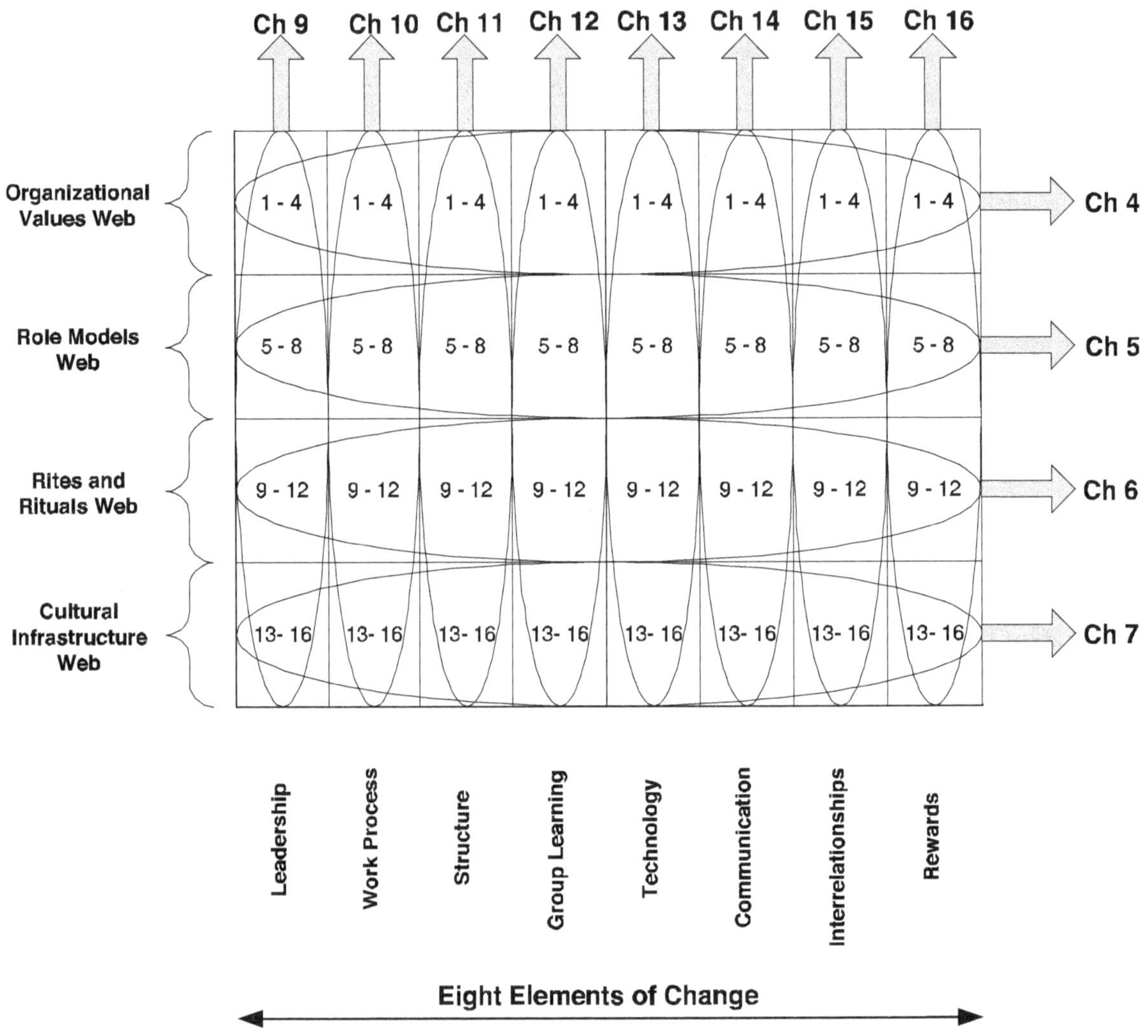

The diagram shows all of the surveys we have taken in this workbook. Chapters 4 through 7 were designed to obtain scores for each of the eight elements of change sorted by each of the four elements of culture. Chapters 9 through 16 obtained scores for each of the four elements of culture sorted by each of the eight elements of change.

Exercise 17-3 Your Web

Part 1

From Chapters 4 through 16, extract the scores from the various web-related exercises and enter their respective blocks in the table below. Chapters 4 through 7 will give you scores for the elements of culture broken down by the eight elements of change. Chapters 9 through 16 will give you the total score for that element. Enter the total score for the element in the block at the bottom of the chart. The total scores from Chapters 4 through 7 should equal the total element scores from Chapters 9 through 16.

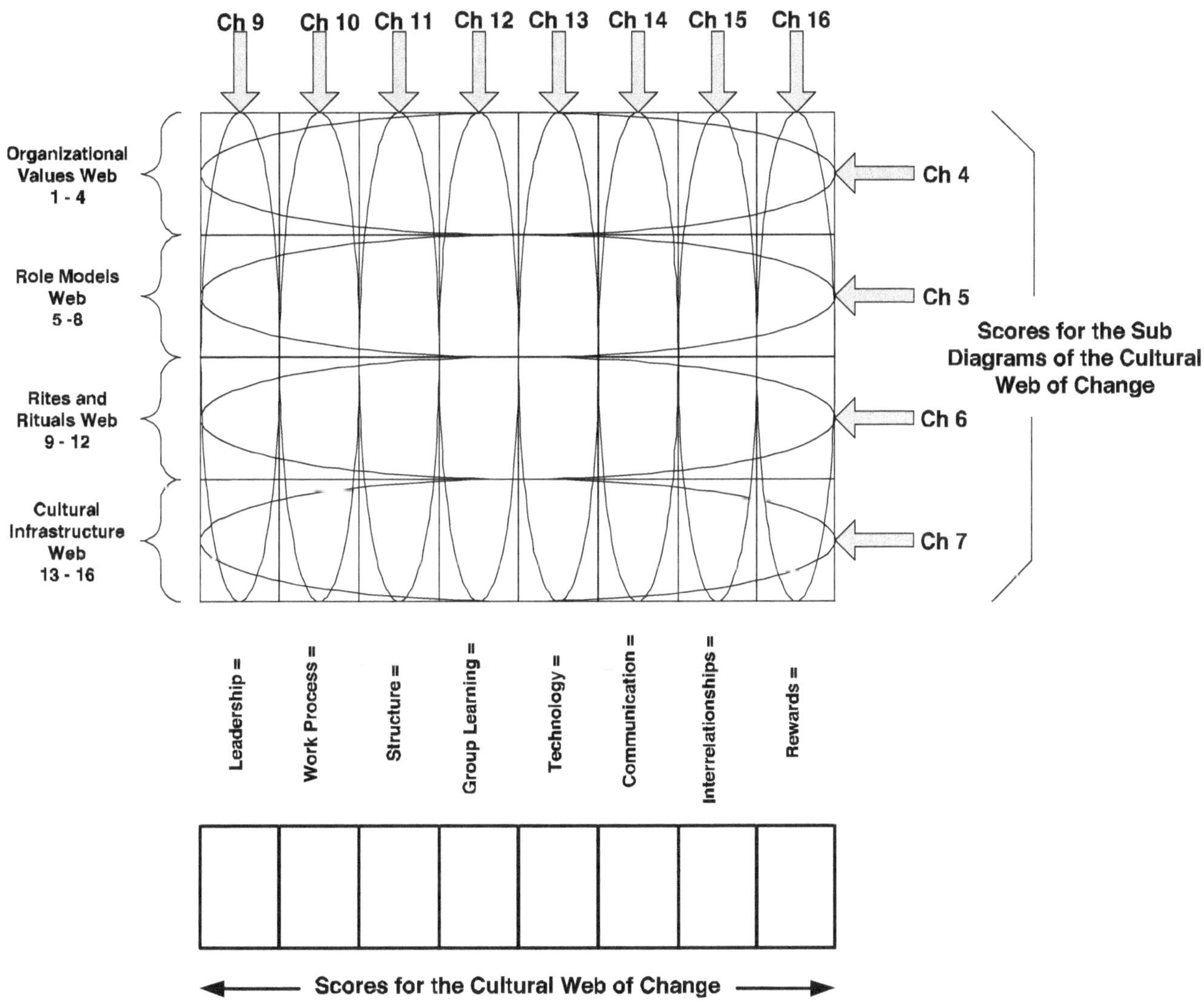

Average Score = the sum of the scores for the cultural web divided by 8 = ______

Part 2

Draw Your Web of Cultural Change.

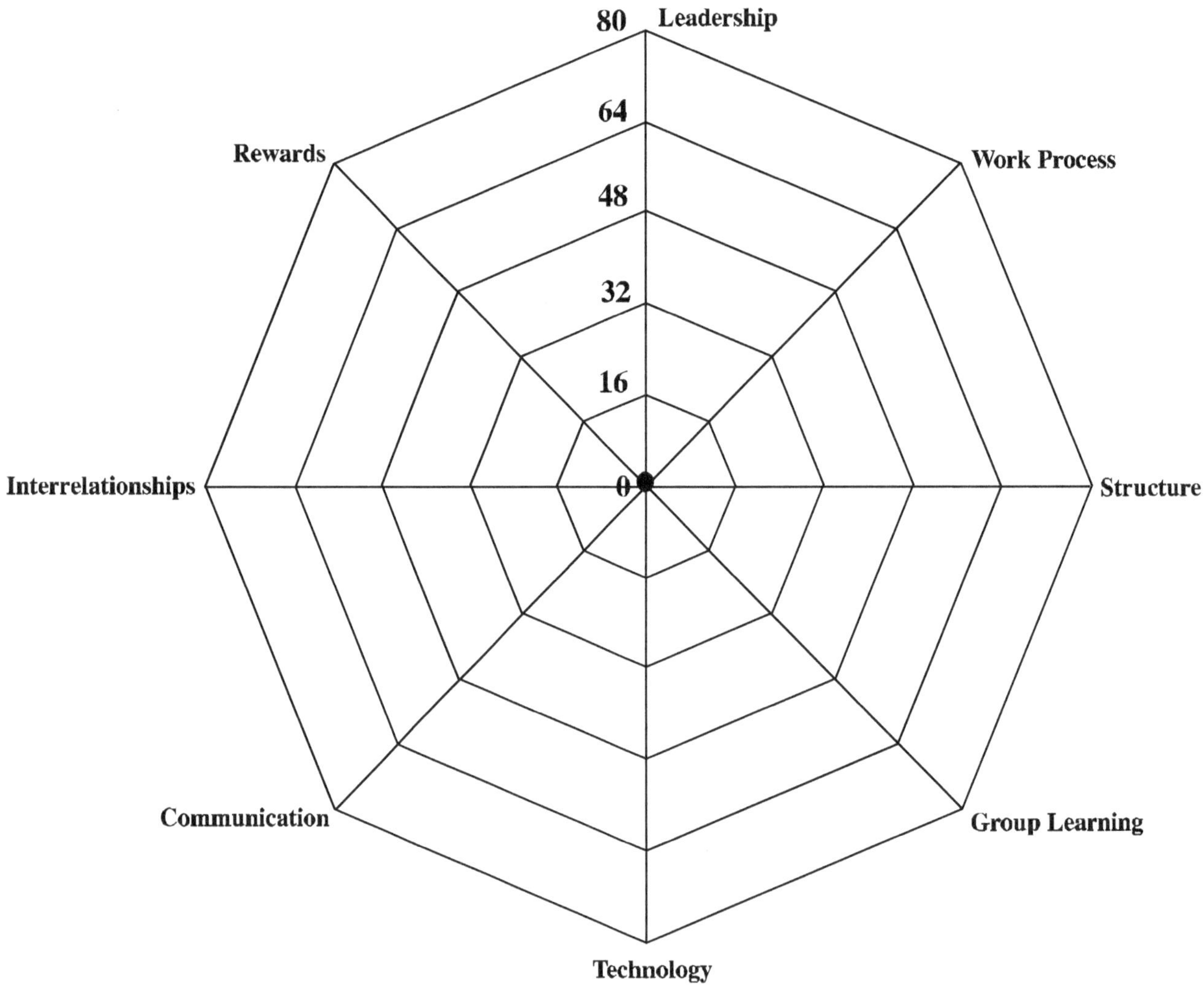

Part 3

Draw Your Sub Webs.

Organizational Values

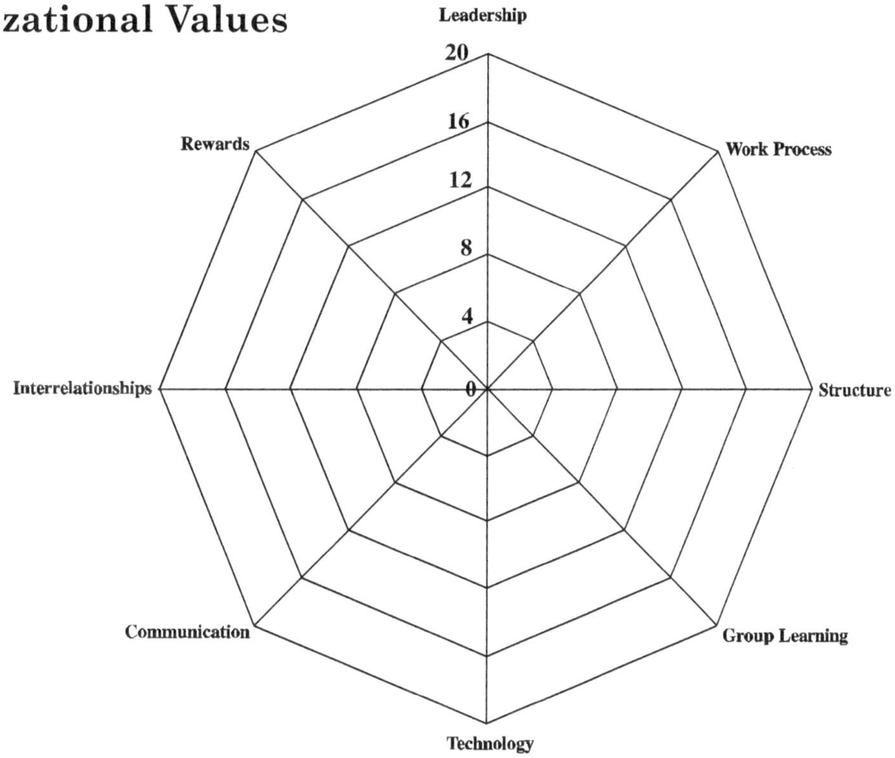

Role Models

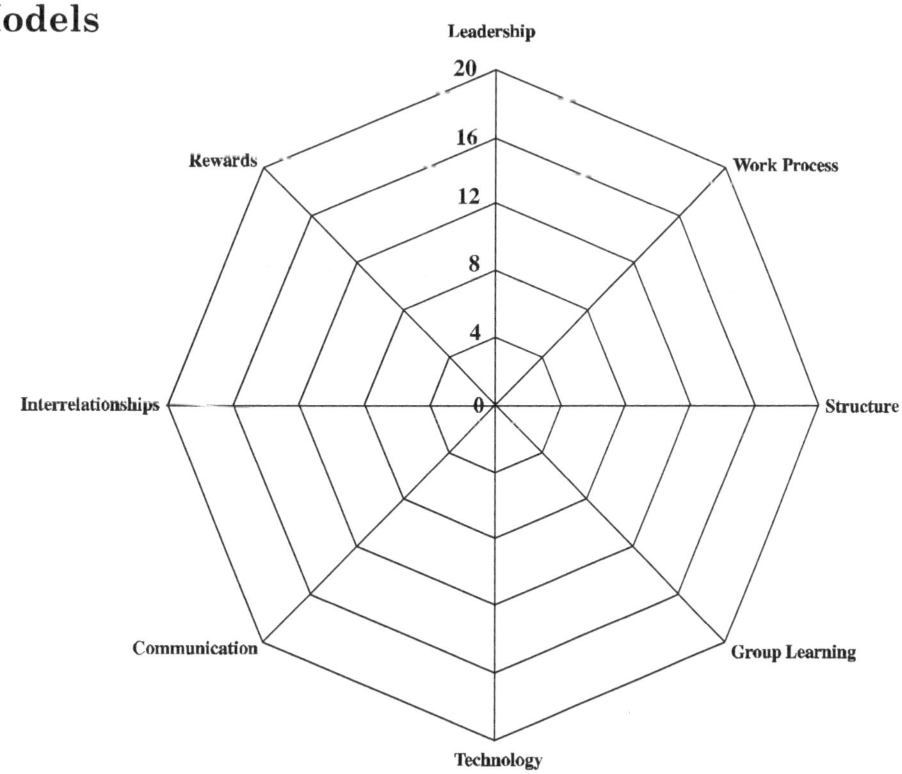

Rituals / Rites

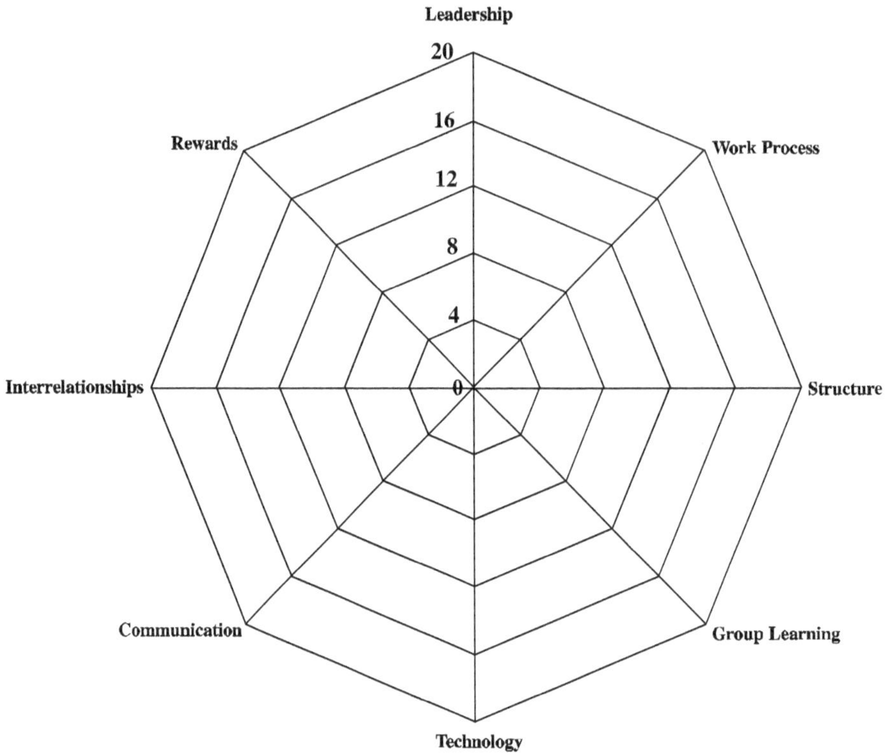

Cultural Infrastructure

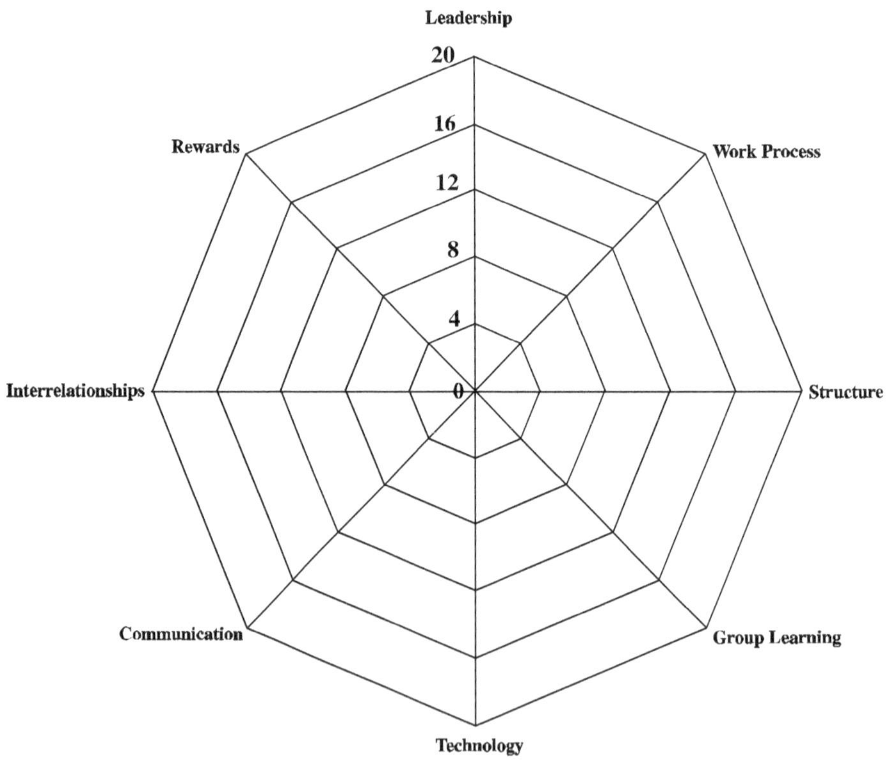

Exercise 17-4 What Are Your Web Diagrams Telling You?

Cultural Web of Change - Analysis of the Full Web

Scores: (0 to 16 very poor, 17 to 32 poor, 33 to 48 average, 49 to 64 good and 65 to 80 very good)

What was the average score and in what range did it fall?
Which of the eight elements of change scored poor or very poor?
Which elements scored in the average range?
Which elements scored good or very good?
How could the poor and very poor scored elements cause problems with your change process?

Organizational Values

Scores: (0 to 4 very poor, 5 to 8 poor, 9 to 12 average, 13 to 16 good and 17 to 20 very good)

Which of the eight elements of change scored poor or very poor?
Which elements scored average?
Which elements scored good or very good?
How could the poor and very poor scored elements cause problems with your change process?

Roles Models

Scores: (0 to 4 very poor, 5 to 8 poor, 9 to 12 average, 13 to 16 good and 17 to 20 very good)

Which of the eight elements of change scored poor or very poor?
Which elements scored average?
Which elements scored good or very good?
How could the poor and very poor scored elements cause problems with your change process?

Rites and Rituals

Scores: (0 to 4 very poor, 5 to 8 poor, 9 to 12 average, 13 to 16 good and 17 to 20 very good)

Which of the eight elements of change scored poor or very poor?
Which elements scored average?
Which elements scored good or very good?
How could the poor and very poor scored elements cause problems with your change process?

Cultural Infrastructure

Scores: (0 to 4 very poor, 5 to 8 poor, 9 to 12 average, 13 to 16 good and 17 to 20 very good)

Which of the eight elements of change scored poor or very poor?
Which elements scored average?
Which elements scored good or very good?
How could the poor and very poor scored elements cause problems with your change process?

Assessment and Corrective Action

Chapter 18

Purpose

This chapter shows you how to take the next step from creating your web of cultural change based on the survey scores to web analysis and, ultimately, corrective action. After all, simply surveying and identifying the gaps is a worthless exercise if you do not do something about closing the gaps.

What Is Included

Figure 18-1 Assessment

Figure 18-2 Technique for Analysis

Figure 18-3 A Sample C-RCFA Diagram

Figure 18-4 C-RCFA Rules

Exercise 18-1 Your C-RCFA

Figure 18-1 Assessment

Change–Root Cause Failure Analysis (C-RCFA)

In this model, you identify the problem. Then by asking why the problem exists, you drill down through several layers to reach the root of your problem. Once you get to this point, you identify and implement solutions to solve the root cause. Taking these steps ultimately helps you solve the problem. A single root cause is often an oversimplified answer; most of our issues have more than one root. However, if you know the methodology associated with a simple solution, you'll find it only takes a little more work to extrapolate the process to a problem with more than one root.

Figure 18-2 Technique for Analysis

Step 1 – The Question and Level 1 Answers

Step 2 – Level 2 Answers

Step 3 – Level 3 Answers

Step 4 – Analysis

Step 5 – Fix the Root / Fix the Problem.

Figure 18-3 A Sample C-RCFA Diagram

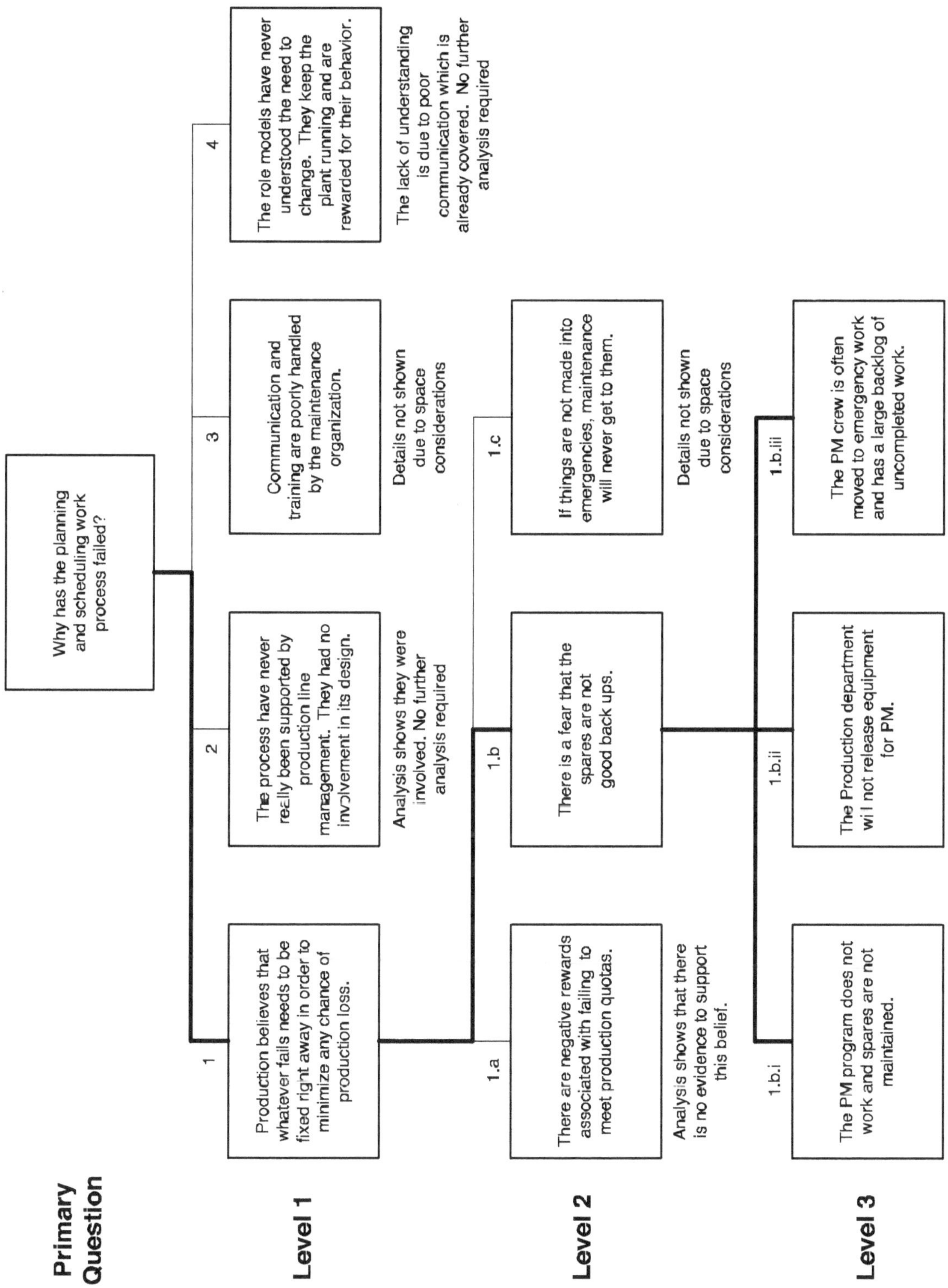

Figure 18-4 C-RCFA Rules

1. Whether the element you work on is one of change or one of culture, select the one with the lowest web score to begin. Solving the root cause of this low score will probably provide you with the most value. You can work on others later.

2. The initial question should be: Why did the _________ element score so low?

3. Don't restrict your answers. You will validate or invalidate them after they are all listed.

4. Write all answers down on the sheet provided as a statement of fact. You will then be able to ask why at subsequent levels of the C-RCFA diagram.

5. After you develop the list answering the question for a specific level, critically determine if the answers are valid for the next level.

6. For each of the valid answers in the prior level, ask why, then write the answer again as a statement of fact.

7. Continue working down through at least three levels.

Exercise 18-1 Your C-RCFA

The exercise provided is designed to help you learn how to conduct a C-RCFA for any of the eight elements of change or four elements of culture that scored low on your web diagrams. To analyze how any one of the eight elements of change impact one of the four elements of culture, refer to the surveys in one of the following chapters;

Elements of Culture	Chapter
Organizational Values	4
Role Models	5
Rites and Rituals	6
Cultural Infrastructure	7

To analyze how any one of the four elements of culture impact one of the eight elements of change, refer to the surveys in one of the following chapters;

Elements of Change	Chapter
Leadership	9
Work Process	10
Structure	11
Group Learning	12
Technology	13
Communication	14
Interrelationships	15
Rewards	16

Step 1 – The Question and Level 1 Answers

The initial question:

Why did the element score so low?

From your web of change diagram and scores, select the element (radial spoke) that scored the lowest.

Possible Level 1 Answers to the Question: State the answers as facts.

#	**Answer (be as specific as possible)**	**Valid (yes / no)**
1		
2		
3		
4		
5		
6		

After you have brainstormed all possible answers (there may be more than 6 so you may need more paper), go back and evaluate whether or not the answers have validity. Those that do will be used at Level 2 and should be marked with a "Y" for yes. Those that do not will be marked with an "N" for no in the block to the right of the respective answer.

Step 2 – Level 2 Answers

The question at Level 2:

Take each valid answer at Level 1 and ask "Why?"

Possible Level 2 Answers Enter the answer number from level 1 in the left column marked L1. Use only those L1 answers that you determined to be valid.

L1 #	L2 #	Answer (be as specific as possible)	Valid (yes / no)
	1		
	2		
	3		
	4		

L1 #	L2 #	Answer (be as specific as possible)	Valid (yes / no)
	1		
	2		
	3		
	4		

L1 #	L2 #	Answer (be as specific as possible)	Valid (yes / no)
	1		
	2		
	3		
	4		

L1 #	L2 #	Answer (be as specific as possible)	Valid (yes / no)
	1		
	2		
	3		
	4		

After you have brainstormed all possible answers (there may be more than 4 for each Level 1 valid statement), go back and evaluate whether or not the answers at level 2 are valid. Those that do will be used at level 3 and should be marked with a "Y" for yes. Those that do not will be marked with an "N" for no in the block to the right of the respective answer.

Step 3 – Level 3 Answers

The question at Level 3:

Take each valid answer at Level 2 and ask "Why?"

Possible Level 3 Answers: Enter the answer number from Level 2 in the left columns marked L1 and L2. Use only those that you determined to be valid.

L1 #	L2 #	L3 #	Answer (be as specific as possible)	Valid (yes / no)
		1		
		2		
		3		
		4		

L1 #	L2 #	L3 #	Answer (be as specific as possible)	Valid (yes / no)
		1		
		2		
		3		
		4		

L1 #	L2 #	L3 #	Answer (be as specific as possible)	Valid (yes / no)
		1		
		2		
		3		
		4		

L1 #	L2 #	L3 #	Answer (be as specific as possible)	Valid (yes / no)
		1		
		2		
		3		
		4		

L1 #	L2 #	L3 #	Answer (be as specific as possible)	Valid (yes / no)
		1		
		2		
		3		
		4		

L1 #	L2 #	L3 #	Answer (be as specific as possible)	Valid (yes / no)
		1		
		2		
		3		
		4		

Note: You can drill down deeper and identify Level 4 answers if needed.

After you have brainstormed all possible answers, go back and evaluate whether or not the answers at Level 3 are valid. Those that do should be marked with a "Y" for yes. Those that do not will be marked with an "N" for no in the block to the right of the respective answer.

Step 4 – Analysis

The next step is to create a C-RCFA diagram similar to the diagram shown in Figure 18-3. Once it is drawn and the blocks filled in with your Level 1, 2 and 3 answers, determine which of the Level 3 answers is the most likely cause of the low element score. By solving this problem, you should be able to trace the corrective action back up through the levels of the C-RCFA diagram and see how it will serve to resolve the low element score. Note that you may have to work on more than one of the Level 3 answers, but at least you will be working on making correction to the root of the problem.

Step 5 – Fix the Root / Fix the Problem.

A suggestion for corrective action is to employ the Goal Achievement Model so that the initiative and the related activities are clearly defined and worked to completion. This information is included in both of my books: Successfully Managing Change in Organizations: *A Users Guide and Improving Maintenance and Reliability Through Cultural Change,* as well as in this workbook.

Chapter 19

Moving Forward

Purpose

This chapter wraps up the work that was done in Chapters 1 through 18 by addressing two factors that are critical for success: readiness and sustainability.

What Is Included

Figure 19-1	Readiness
Figure 19-2	Steps in the Readiness for Change Process
Exercise 19-1	Your Readiness Approach
Figure 19-3	Sustainability
Figure 19-4	Sustainability Tools
Figure 19-5	Your Learning Spiral
Exercise 19-2	Your Work Plan
Figure 19-6	Final Thoughts

Figure 19-1 Readiness

"You don't have to be ready to begin the process of change. Many initiatives are started without those involved being ready.
They usually have one thing in common – failure."

Steve Thomas 2001

Figure 19-2 Steps in the Readiness-for-Change Process

- **Develop a clear vision of the future.**
- **Communicate the vision.**
- **Involvement – understand next steps.**
- **Training**
- **The leaders must "walk the talk."**

Exercise 19-1 Your Readiness Approach

Using the readiness steps listed in Figure 19-2, think of a change initiative in which you have been involved. What were the positive or negative aspects of your readiness process related to each of the steps? What was missed? What was the result?

Develop a clear vision of the future.

What were the positive or negative aspects of your readiness process?
What was the result?

Communicate the Vision

What were the positive or negative aspects of your readiness process?
What was the result?

Involvement – Understand the Next Steps.

What were the positive or negative aspects of your readiness process?
What was the result?

Training

What were the positive or negative aspects of your readiness process?
What was the result?

Leaders Must "Walk the Talk."

What were the positive or negative aspects of your readiness process?
What was the result?

Figure 19-3 Sustainability

Sustainability is the ability to integrate the process changes into the organization's culture so that it will not be easily changed due to changes in the organization. The change, once implemented, needs to be viewed not as additional work, but rather as how we work!

Figure 19-4 Sustainability Tools

- Communicate the vision, work status, and other things of importance.
- Use the Goal Achievement Model
- Use the Roadmap of Change
- Specific roles and responsibilities
- Use of the Web of Change and Web of Cultural Change
- Ongoing audits with action items
- An aligned reward system
- Training for new employees and follow up
- Refresher training for the organization
- Continuous improvement process using group learning principles

Figure 19-5 Your Learning Spiral

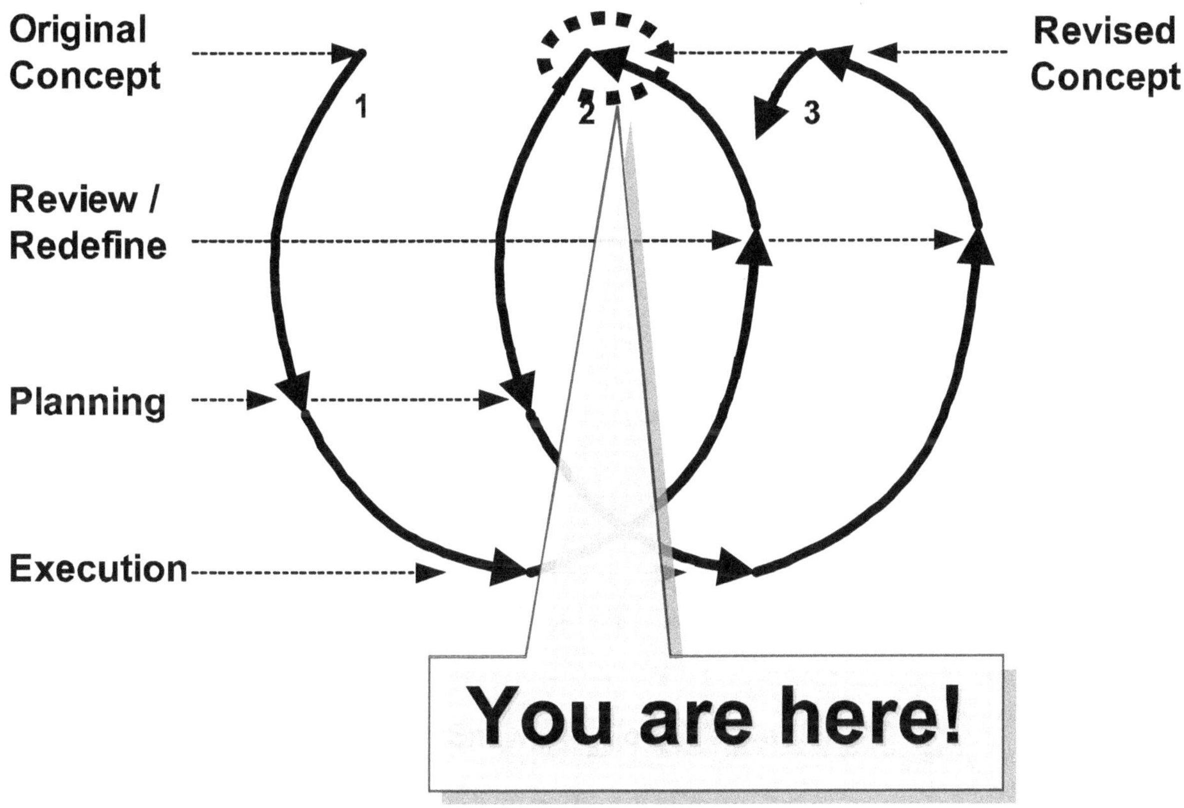

So what are you going to do with your newly acquired information?

Exercise 19-2 Your Work Plan

What are your plans to use the information that you have learned to improve reliability and maintenance within your plant or company? List several initiatives that you will develop and work on when you return.

Initiative #1
Initiative #2
Initiative #3
Initiative #4

Figure 19-6 Final Thoughts

- Cultural change is an essential part of what we do and who we are.
- Without change, we stagnate and die – as companies and as individuals.
- With change, we are able to constantly renew ourselves and continuously improve.
- Change is not easy, but it is always the right thing to do.
- Change takes persistence, commitment, and a good sense of timing.
- Change requires an overriding desire to do what is right, often in the face of seemingly insurmountable odds.
- Your motto needs to be "never give up trying" because, once you do, all is lost.

If you embrace these thoughts, one day you will find you have reached the mountain top and successfully achieved the changes you sought to implement. And on the horizon I guarantee you will find another mountain because the journey of the change agent never ends.

Steve Thomas

Improving Maintenance and Reliability Through Cultural Change *(2005)*

The American Tractium Production Company

Appendix 1

The Company

The American Tractium Production Company, or ATPCo, was founded 25 years ago by president Allen Peters. It manufactures tractium, a metal-like material that results from a manufacturing process that takes metallic scrap and, through a conversion process, creates a durable material used to create many goods for the home.

ATPCo takes in raw material, creates tractium, and then manufactures tractium products. Sales are all through large chain stores that sell goods for home use. ATPCo does no direct marketing.

Because the manufacturing process uses other companies' scrap material, supply is virtually unlimited. ATPCo has entered into several long-term contracts that will supply their processes in excess of the current and foreseeable needs. Companies are happy to sell ATPCo their scrap as an alternative to costly scrap disposal.

After the initial discovery, it was found that there were other processes to create tractium. As a result, other firms now use different processes to produce competitive goods. One factor that limits competition, however, is the high capital cost of entry into the market. The equipment used in both the tractium conversion and manufacturing processes is extremely expensive. Firms currently in the market can supply both current and projected future needs. Therefore, there is little interest by nonparticipants to enter. Because the finished goods are common household devices, there is little brand recognition and little company loyalty. If ATPCo does not get its goods into the large stores selling their product, stores will buy from ATPCo's competition.

ATPCo has five plants located strategically across the country. They are the Eastern, Northern, Southern, Western, and Central plants. Because raw material is widely available, shipping it to the plants is not an issue. ATPCo also built its plants to be close to its major markets, minimizing transportation costs of the finished goods. This further allows ATPCo to respond to its customer's needs at a regional level. This strategy has enabled ATPCo to keep its production costs low and its competitive position high within the industry.

Each plant is essentially the same. The process is a single train. Facilities receive the raw material, prepare it for conversion, produce the actual tractium, and then manufacture finished goods. The manufacturing process is altered based on the finished goods being produced. With the desire for a high degree of manufacturing flexibility, each site has been built with more than one processing line. The size of the facility is based on regional demand plus additional unused capacity for future growth or short-term capacity increases. The manufacturing process is shown in Figure A-1.

Corporate

The vision for the company is as follows: Operate the facilities in a manner that optimizes production and is capable of effectively and efficiently meeting the demand of the market.

Each plant is a self-sufficient entity with all the departments necessary to operate its business locally. Company president Allen Peters believes in site self-sufficiency; he maintains only a small corporate staff. Within each plant, department managers report directly to their site manager; they have dotted-line reporting to a corporate manager with the same functional responsibility. Thus each site is self-sufficient, yet the corporation coordinates each functional area. The corporation provides cooperation, coordination, and sharing of items of corporate or common interest.

It is expected that the plants work together, share information, and help each other out. For example, if one plant has production problems and can not deliver product, it is expected that another will increase production to fill the void. This solution is short-term because shipping across regions increases costs and reduces profit.

The Workforce

The workforce in each plant is non-union. As a result, there are few labor relations issues. A working committee of plant management personnel and representatives selected by the workforce resolves problems that do arise. The majority of the plant employees are either the operators who run the tractium lines or the maintenance mechanics responsible for fixing the equipment when it breaks. The other organizations have small staffs to support their individual functions.

The Product and Sales

At present, there are four products made by ATPCo and its competitors. They are referred to as Alpha, Beta, Cappa, and Delta. The final manufacturing process needs to retool its equipment each time a different product is put on their line. Fortunately, this retooling process is simple; it is accomplished with little effort or downtime. As a result, the production process is based on whatever Sales can sell. As Sales obtains orders, the manufacturing lines are scheduled to produce that order.

The Eastern Plant

The Eastern plant is like all of the others in design. Its distinctive features is its size. It has five independent production lines. It is also the newest and most

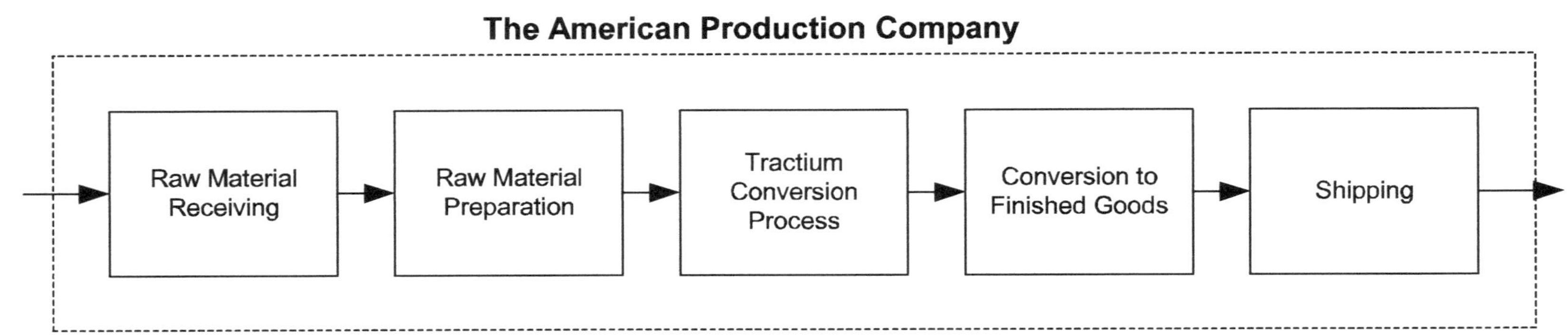

Figure A-1

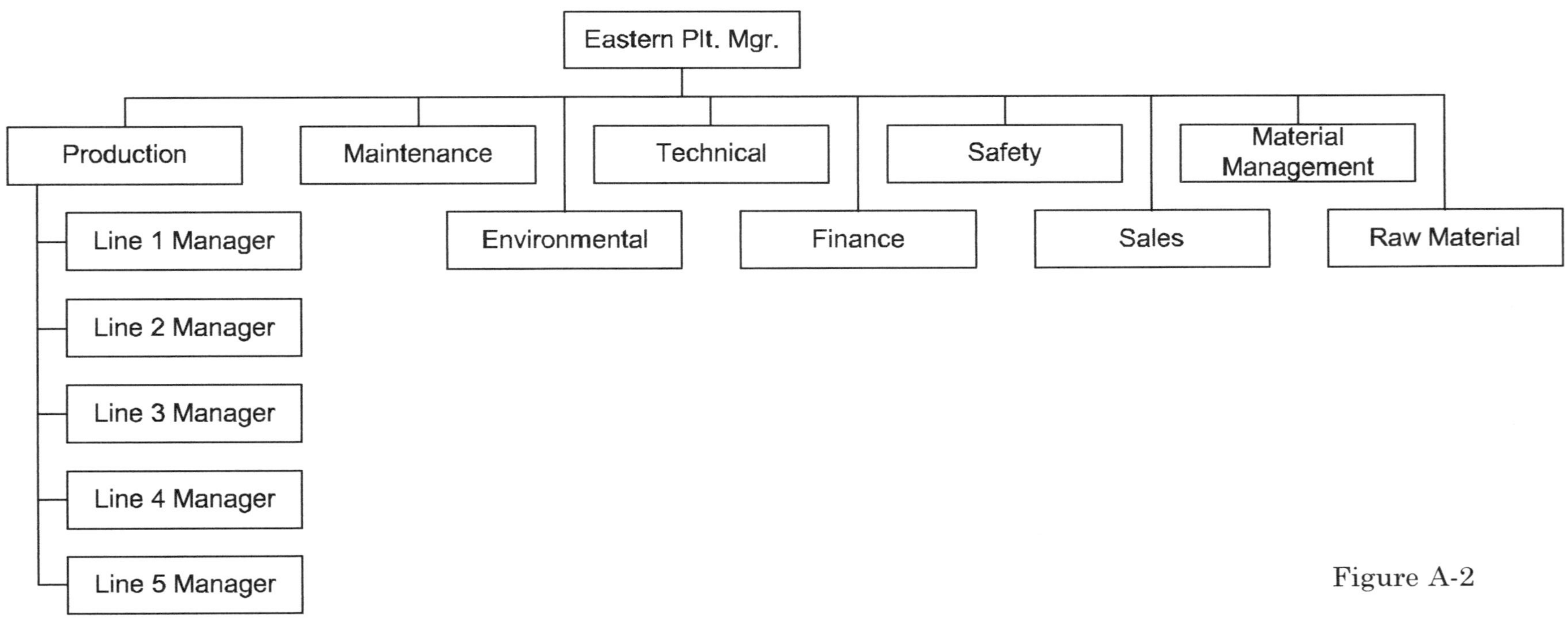

Figure A-2

recently built plant. ATPCo constructed it ten years ago to help to complete its nationwide expansion. Started on the West Coast, ATPCo had been expanding eastward for 13 years. Finally it acquired the right property for its eastern plant. Two years later the new plant went into production.

The plant's organizational structure, which follows the corporate model, is shown in Figure A-2.

The Eastern Plant – Organizational Detail

Production

Each of the five lines in the plant has a line manager and a group of production line workers. These workers, who comprise a crew, are hourly employees. When the Eastern plant first opened, it had only a day crew working Monday through Friday. As you will see, production demands have required that crews now work 12-hour shifts seven days per week. The organization of a production line shown in Figure A-3.

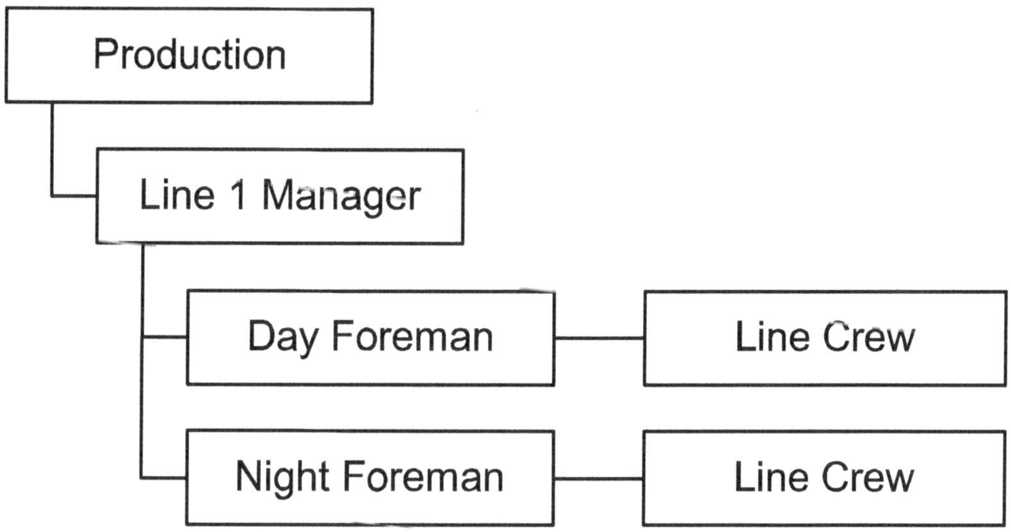

Figure A-3

Maintenance

In the maintenance department, the foremen and mechanics are multi-skilled. They can perform mechanical, machinery, and instrumentation tasks equally well.

The department also has a separate instrument group for the highly-specialized controls, as well as a separate electrical group. Three general foremen each have a crew of 10 mechanics. They are dispatched to the jobs by a work scheduler. They have no assignment to a specific line. The special instrument crew and the electrical crew are dispatched in a similar manner. The maintenance organization believes that working in this manner allows them to optimize the resources they can commit to any specific job. There are also line mechanics, two for each line. They are highly-skilled individuals who work with the line crew to keep the line running. The line has one mechanic assigned to each shift.

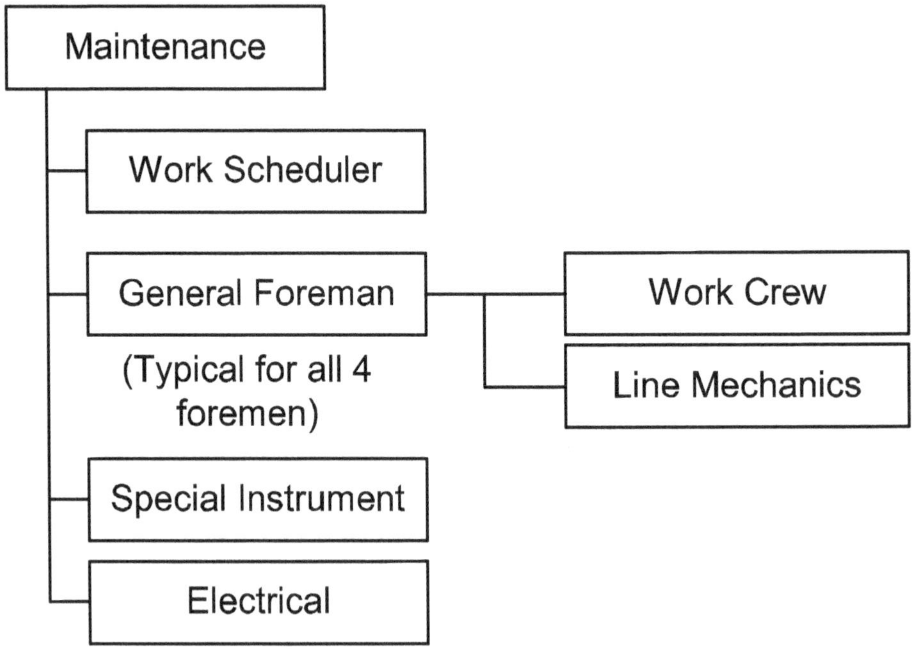

Figure A-4

Technical

The technical group's employees are all of the engineers, including: rotating equipment, instrument, and electrical engineers as well as a generalist category referred to as line engineers. One line engineer is assigned to each of the five production lines. The specialty engineers are assigned from a central group. This approach follows the same philosophy as that of the maintenance organization, allowing the group freedom to apply the correct resource to any particular problem.

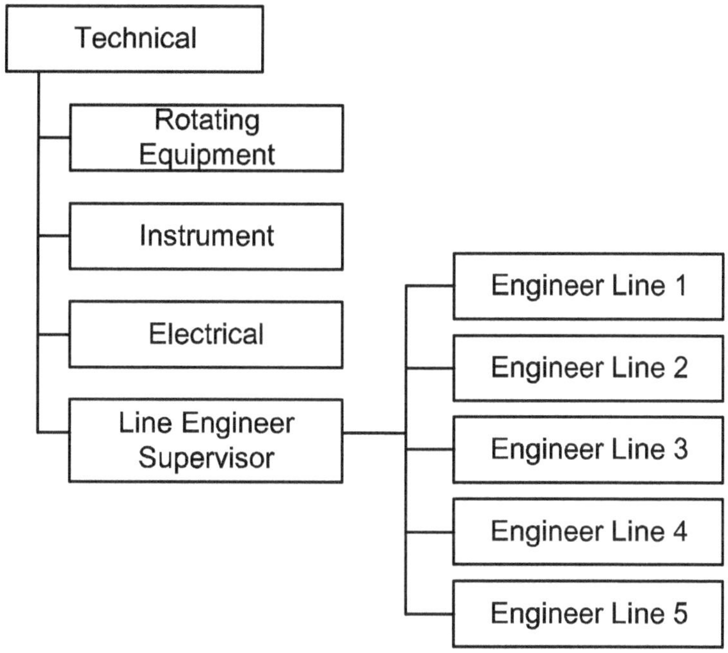

Figure A-5

Leadership

The site manager, Mike Kane, has been with the company since it was created. He and Allen Peters have worked side-by-side to create a highly productive, and very competitive firm. During the early years, the company struggled to survive. Tractium was an unknown product and the production process was not mature. In the early days, the goal was not to optimize production, but "maximize production," simply because the process was not efficient. It was subject to numerous breakdowns. The company had to maximize the production runs just to satisfy the orders.

Mike Kane was the right person to work out these problems. He was a focused, hard-driving manager. He drove the equipment and the people, and got results. His most frequently used quote was "my way or the highway." As the business grew, his style of managing no longer applied. The equipment was efficient and had excess capacity. Optimization became the company direction. However, Allen Peters felt a high degree of loyalty to Kane from all of his early work. Consequently he assigned Kane to the Eastern plant. Allen Peters believed that because this plant was new, modern, and highly reliable, Kane could finish out his career in virtual ease.

Production Strategy

The production strategy is to manufacture products Alpha, Beta, Cappa, or Delta to satisfy sales orders. Because the Eastern plant has five lines, and retooling is an easy process, the lines are not at capacity. The production lines operate eight hours per day, Monday through Friday. Machine downtime is not an issue; the production rate can be raised or the lines run during the other shifts to make up for any time a line is shutdown. This flexibility results from designing the plant with excess capacity. However, operating during other shifts or the weekend to make up for lost production is not Kane's style. He prides himself on being able to rapidly repair whatever breaks and return the line to service. He then pushes employees and the equipment to raise the rate, making up lost production without overtime or off-shift work. Kane believes that other alternatives are not productive; they reflect negatively on his "can do" operating principles.

Maintenance Strategy

When the plant was relatively new, the equipment was very reliable. Breakdowns were rare. Coupled with Kane's strategy of rapid repair, the plant had achieved two things. First, the production lines had highly-skilled mechanics assigned to work with the line crews. Their job was to fix equipment as quickly as possible, returning the production line to service. Over time they got very good at this. The on-site warehouse stocked whatever parts they needed and the mechanics were trained in rapid-repair techniques. One of these techniques was to have replacement equipment and parts in a standby mode. Then when equipment failed, a quick change out was executed, resulting in minimal downtime. The second achievement was that the centrally-dispatched maintenance organization also became very good at rapid repair. With few failures, due to the newness of the equipment and the rapid replacement strategy, they essentially worked under the premise of "when it breaks, we fix it."

As the plant passes its tenth year of operation, the equipment is beginning to show its age. The reliability that was once present is no longer there. As a result, the mechanics have to work harder to service failed equipment and maintain production. Kane's strategy of raising the rate to make up lost production works in the short term. However, over the long term, this strategy causes the equipment to deteriorate even further. The result is increased failure, higher repair costs (including additional spares), and increased cost of production.

The decreasing reliability of the equipment and the frequent need to work both middle and late shifts to achieve the production quotas has Kane in a perpetual state of frustration and anger.

A Major Change for ATPCo

Just recently a new use was discovered for tractium. The new product Epsilon, is in great demand. This has caused several things to happen that have had a serious impact on ATPCo. First, the potential profit from Epsilon has justified the cost for new companies to enter the market. In turn, their entry has driven up the price of new equipment. Furthermore, delivery time for new equipment has lengthened, making any expansion by ATPCo relatively difficult in the short term. Fortunately, ATPCo had built its plants with excess capacity; therefore, it had an initial advantage over its competition. ATPCo had already started both a middle and a late shift on a part-time basis to make up for the increased downtime that resulted from equipment reliability issues. These shifts have since been expanded to full-time operation. Thus, ATPCo has increased production and taken advantage of the increased marketability of both existing and new tractium products.

The other major problem ATPCo has encountered was one that was becoming obvious even before the new markets opened. Equipment reliability has been deteriorating. In the new manufacturing environment, there is no spare capacity. Three shifts are already working to produce the required product, and the cost of production is increasing due to equipment failure. Nowhere is this felt more strongly than in the Eastern plant where the new market is geographically centered.

The Eastern Plant Problem

Allen Peters has a big problem. He has a new, highly-competitive environment in which ATPCo must operate. He needs to make not only the existing products that ATPCo has made for years, but also the new product Epsilon. It is in such demand that the Eastern plant is now running two of its five lines 24 hours per day, seven days per week to meet the sales orders. The other three lines are also running full-time to produce Alpha, Beta, Cappa, and Delta. ATPCo can't expand in the short-term because new firms have bought all available or soon-to-be-manufactured equipment from its suppliers. The profit margin for Epsilon is small; shipping it from other sites is not a practical solution. Finally Mike Kane, his Eastern plant manager and friend only knows one way to operate—drive the equipment hard and fix whatever breaks quickly. Allen Peters knows this is the wrong approach. He wants a manager who will create a reliability-based environment. He needs his equipment to operate all of the time, with unplanned failures never occurring. This reliability-based strategy is the only way he sees that ATPCo will survive.

www.ingramcontent.com/pod-product-compliance
Lightning Source LLC
LaVergne TN
LVHW080013210926
848481LV00011B/3015